IT ALL CAME FROM NOTHING

THE INEVITABLE UNIVERSE

Contents

Prelude

"Behind it all, Is surely an idea,
So simple, So beautiful, So compelling,
That when we grasp it,
We will all say to each other,
How could it have been otherwise."

John Archibald Wheeler

The zero energy cosmos comes with enabling initial conditions. A void devoid of time and space, is unstable, Expansion *ex nihilo* is inevitable for the same reason it is regenerative.

The physics of *emptiness* was first studied by the Dutch physicists, Willem De Sitter in 1917. What is now believed to be the real universe, moderates positive and negative energies equally to maintain perfect balance. Like the empty universe, the present cosmos is believed to be intrinsically flat and infinite (or potentially infinite) in extent.[1] Conflation creates negative pressure, expansion of negative pressure creates positive energy. The former takes form as tension (spatial stress), the latter appears initially as radiation. From spatial stress, particles emerge with inertial properties. Mass forged from expansion of negative pressure is equalized by the expanding volume of negative pressure. Whatever be the truth as to how it all began, the present state of the universe is matter dominated. The primary form of negative energy now exists as the gravitational fields of particles.

To model the *"now"* state of our evolving universe requires a knowledge of its history and an understanding of the interplay between gravity and expansion. To look into the past, we peer deep into space, only to see our own puzzled faces looking back in wonderment. That which provokes expansion on the large scale, also creates the attraction between masses on a lesser scale.

Gravity is weak by comparison with other forces. But was it always so? As developed herein, gravity depends upon both the stress-energy density of space and the cosmological expansion rate. The reader having some familiarity with cosmology will immediately object that gravity is curvature, and to understand the geometry of bent space and altered time, one must study General Relativity. The problem is, GR does not explain how or why mass-energy conditions spacetime. Like Newton's Gravity Law, GR is a calculation tool. Theories are limited by the knowledge available at the time of their creation. To demystify gravity and inertia, it will be necessary to combine the discoveries of the 20th century in a new concatenation.

Introduced descriptively using only words and diagrams, Gravity conceptualizes as an amalgamation of two physiol-mathematical principles. Then by mathematical means, formulations are derived from which the theory may be tested, accredited, reproved or falsified.

1) Gravity is the inertial reaction of individual particles brought about by spatial expansion, and

2) The dynamic impedance of flat space manifests as a plurality of operative infinite planes.

---

[1] A curious prediction of Einstein's field equation is that an empty universe (one devoid of both pressure and matter) naturally expands exponentially. The same fate follows when negative and positive energies are balanced..

The Hubble sphere, in its function as an observational limit, offers nil resistance to acceleration in its form as a three dimensional density. How is it that nearly empty space impedes changing motion? Velocities are respective to one another, mensurable by the difference in the rate at which time is accumulated by relatively moving clocks. In contrast, changing velocity is relative to the universe as a whole. Accelerations create inertial forces proportional to the masses. Traditionally, the mystery of inertia has been reduced to one of two alternatives: Either:

1) Mass-energy is intrinsically endowed with an acceleration impeding quality, or,

2) Matter and space coalesce as a unified global inertial property of the cosmos.

Historically, the idea of inertial reaction as dependent upon the other matter in the universe, can be traced back to the early Greek philosopher, Anaxagoras: *"In everything there is a portion of everything."*[2] The objection to the organic theory of inertia has always been that of 'time.' Einstein reasoned, a mass at rest would feel a reaction force if the entire universe were accelerated. But he took issue with how distant matter could make its presence instantly felt by local bodies? Like others before and after him, he brooded over what is now called *"Mach's Principle".*[3] But today's cosmologist has an advantage - being now informed in such a way as to believe: Space is geometrically flat and potentially infinite. Discoveries near the end of the 20th century paint a picture of the universe that will accommodate the operative geometry of an infinite plane. Inertial reaction can now be explained and defined in terms of Hubble parameters, not as a 3-D sphere, but rather as a functional ensemble of infinite planes. Crudely speaking, the physiology of any single plane can be imagined as the positive bare mass energy contained in a Hubble sized sample of the universe, spread uniformly over a flat plane having the same surface area as the Hubble sphere. The surface density -- by the principle of cosmic homogeneity, will be the same everywhere in the universe.

Infinite planes have a unique mathematical property. They exert the same gravitational and inertial influence everywhere. Consequently, one infinite plane is mathematically sufficient to specify the operative inertia of the universe. An infinite plane exerts the same gravitational force upon a mass irrespective of its distance from the plane. While the density is determined by a sample size, any size will be satisfactory, so long as it is sufficiently large to obtain a good estimate of average density. As later developed, the area density of a plane formed by spreading the estimated mass of the Hubble sphere over a surface commensurate with its manifold, is in the range of one kilogram per square meter. That it is exactly one kg per square meter is a result of the definitions established between force, mass and acceleration by the early experimenters.

---

[2]In the 18th century, Bishop Berkeley proposed a more specific application of the holistic principle, suggesting inertial reaction to be the result of all other matter in the universe. Ernst Mach put forth a similar theory in the 19th century, which as a consequence of Einstein's interest therein, is now known as Mach's Principle.

[3]Einstein, although having initially considered Mach's principle it the formulation of GR, eventually dismissed it as being in conflict with his theory of Special Relativity. In his 1920 address at Leyden University, Einstein said: *"...inertial resistance opposed to the relative acceleration of distant masses presupposes action at a distance, and as the modern physicist does not believe he may accept action at a distance, he comes back once more to the ether..."*

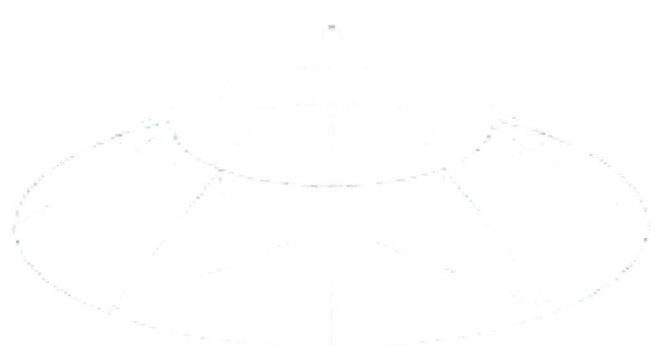

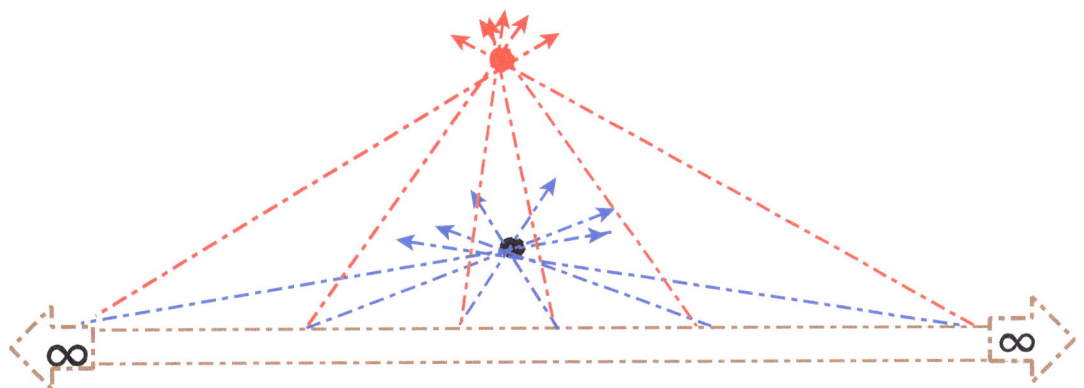

Fig 2: Mass contained in the spherical volume is spread uniformly over the plane created by transformation thereof (brown slab viewed edge-on) to effectuate a dynamic surface density impedance $\sigma_U = \rho_U R/3$. The force acting upon the red and blue spheres will be the same. Although the blue sphere is closer, the angle of the force lines of the more distant contributions to the verticle force, is less favorable. The direction of the force lines are shown for expanding space or for inertial objects (red and blue) accelerated relative to the plane. Gravitational reactive forces are oppositely directed.

Infinite plane topology emerges as the inertial operative in 3-D Euclidean space –> the inevitable consequence of the conservation laws of momentum and energy. There is no law of "conservation of mass of space." In a zero energy universe, conserving mechanisms for momentum and energy are intrinsic. Momentum, not mass, is fundamental. Any change in momentum must be countered instantly by the reaction of inertial space as counter poise between the local and the whole. The means by which the universe marshals its mass to absorb force is depicted in **Fig 1** and **Fig 2**. In the infinite plane model of flat inertial space, cosmic mass is amalgamated within a spatial surface area to create an area density modulus σ_U. The plane so formed is metaphorical, but it produces physical results in the form of an instantaneous spatial impedance to any change in momentum anywhere .

A second known property of space, is that it expands at an accelerating rate $\mathbf{a_n}$. These two factors (σ_U and $\mathbf{a_n}$), determine the gravitational constant **G**. When multiplied together, they define the vacuum pressure **-P** (expansion creates negative pressure).

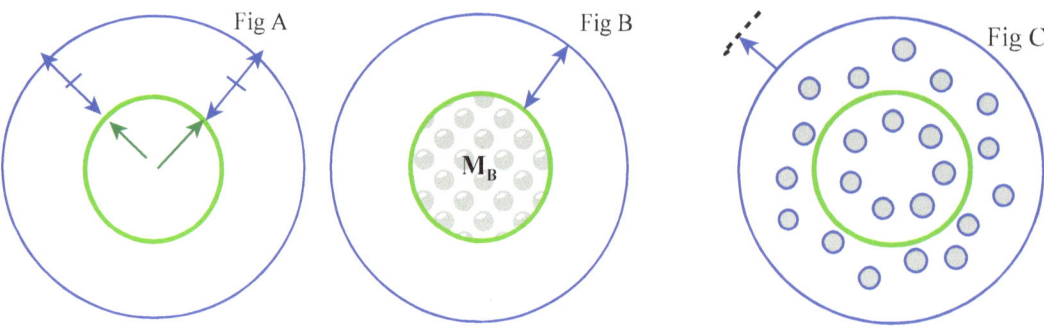

Fig A. Green circle defines an exponentially expanding volume of empty space within a larger volume of exponentially expanding space (blue).

Fig B. The inertial mass of gray spheres populating the green volume impairs spatial expansion. This leads to two possibilities. If the spheres are unbound and sufficiently separated, they will radially diverge with the flow of space from the green volume into the blue volume.

Fig C. Expansion of space within the green volume causes unbound masses to diverge radially outwardly into the empty surrounding blue volume. The dotted black line indicates the spatial volume now occupied by the blue volume. The gray spheres represent galaxies. According to the prevailing theory of the universe, widely separated galaxies co-move with expanding space..

Fig D. The spheres are replaced with tiles to create a shell having the same total mass **M$_B$**. Each tile (gray) is secured to its adjacent tiles by internal forces. Spatial expansion within the tile shell (green arrows within the green volume) produce no physical change (the internal spatial expansion field is too weak to separate tiles from their binding forces). Ensuant, there is no change in the tile shell, or its motion, or its size, nor is their any motion of the individual tiles. That matter is not expanded by expanding space, however, is not without consequence. The inertial property of the non expanding tile shell opposes spatial expansion, flux from the universe (blue arrows) flows inwardly from the universe. Newton called the reaction "*gravity*," but neither Newton nor Einstein recognized the field as an **F = ma** reactionary force. That would be left to Richard Feynman.[4]

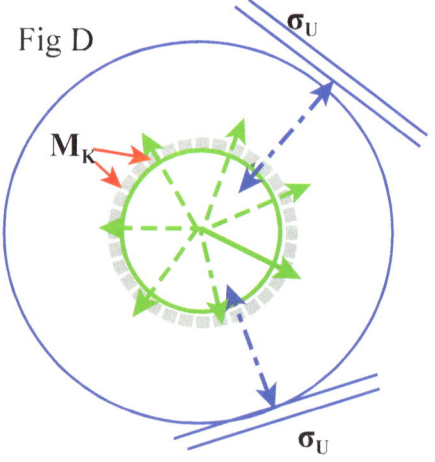

Fig D

---

[4]"One very important feature of pseudo forces is that they are always proportional to the masses. The same is true of gravity. The possibility exists therefore, *that gravity itself is a pseudo force.* Is it not possible that perhaps gravitation is due simply to the fact we do not have the right coordinate system?" [Feynman - Lectures On Physics]

In **Fig D**, the opposition of inertial mass to the free expansion of internal space (green arrows) creates a volume-pressure sink shown absorbing spatial influx (blue arrows) entering the green boundary through the tiles. The magnitude of the influx is determined by the mass of the tiles and the impedance planes σ_U. Net momentum influx created by expansion must equal zero, ergo, the momentum flows must balance on the global scale. As to be later developed, the 'g' fields of individual masses can be explained by spatial influx from the cosmos as well as inertial interference with spatially created outflux (that created by internal spatial expansion) In either case, there is no physical change in the size of the physical objects - expansion does not affect the size of objects Because the totality of cosmic mass acts collectively in the form of an inertial plane, there is definitive autonomy between the Hubble sphere as an entity and the mass of the individual bodies that comprise its content.

The notion of the cosmos as a collective operative defined by an imaginary inertial plane σ_U is indicated by the black arrows (the inertial reaction between an individual tile of mass M_k and the σ_U plane representing the reactionary opposition of the universe). It is understood that M_k can be of any size and located anywhere within the construct of the Hubble volume. To facilitate the verification of field intensity, however, it will prove convenient to consider a spherical mass of uniform density concentric with the surface over which the field strength is to be evaluated.

Fig E: A spherically uniform body of mass M_B surrounded by a concentric spheroidal shell density σ_U. The shell represents a functional composite of sections of **8** infinite planes each operatively orthogonal to the line of action intersecting a point at the geo-center of M_B. The dotted red lines represents a bi-directional bundle of spatial expansion. As to be later developed, a single spherical shell (dotted blue) having an area density σ_U and size commensurate with the Hubble sphere, will have an equivalent inertial influence upon the 'g' field of objects enclosed thereby.

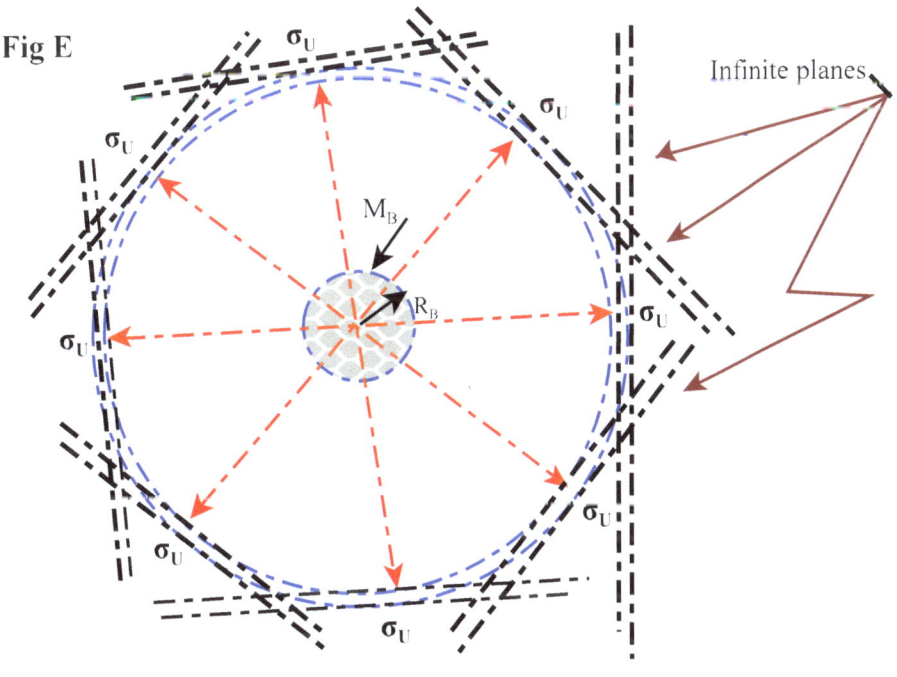

The Hubble sphere defines the observational limit. Consequently it provides a useful sample size for estimating the average density of the larger unobservable universe which may be infinite in extent. As an operative geometry for inertial purposes, however, the 3-sphere volumetric density (ρ_U) fails. It nonetheless provides a proto structure and the physical parameters from which the operative geometry of global inertia can be determined.

While the volumetric density of the Hubble (ρ_U) cannot explain inertial reaction as a 3-sphere construct, it can, by operation of mathematical laws, be transformed to an operative area density exhibiting the same gravitational characteristics (Figure - page 43 - Sidelines on Gravity and Inertia). Then by way of a secondary transformation, the characteristic surface density is preserved when the shell topology is abused to create a flat plane of infinite extent. The ratio formed by dividing Hubble bare mass M_H by its effective surface area ($4\pi R^2$) leads to the same σ_U. That the density determined from the transformation of the Hubble bare mass from 3-sphere to a 2-sphere, also defines the density σ_U of an equivalent plane of infinite extent, leads to an astonishingly simple formulation of gravity. To take advantage of the unique property of the infinite plane (coupling strength is independent of distance), the shell constellation amalgamated from Hubble mass and surface area transformations straight-away to a flat plane of infinite extent. As any appurtenant cosmological reaction must also be three dimensional, ergo the inertial construct of the infinite universe must be operatively isotropic and geometrically convergent (the eight infinite planes of **Fig E** are extended to an infinite number, the cosmic inertia reduces to a single spherical shell of density σ_U

Adverting again to **Fig E**, the force imbalance created by the isotropic spatial expansion field (dotted red bidirectional arrows) acting upon M_B must be taken into account by the universe. From the perspective of the geo-center of M_B, the expansion along any line of action is opposed by the inertial effect of matter. The total reactionary force over a 360 degree span is proportional to the total mass M_B and likewise, the total reactionary impediment of the universe is equal to the mass of the universe M_U which is represented by the spherical surround which would have an area equal to the Hubble surface and consequently and average surface density σ_U. If the interstitial space (that which lies between M_B and the shell M_U) is unimpaired, one might expect that a description of gravity could be written in terms of the masses M_B and M_U. But where gravity is the subject, the interest is in the force density rather than total force. Force density is pressure, and by Pascal's law, pressure is constant throughout a contained volume. It is necessary that spatial pressure created by the reaction of the universe as a whole be equal to the pressure created by the inertial reaction of M_B. Pressure is momentum flow. When the pressures are equal, momentum flows are equal (A requirement of the zero energy universe).

Taking Newton's 2nd law as symmetrical, what is the effect of isolated chunks of matter such as M_B upon the expansion of space? The acceleration of mass wrt space creates $F = ma$ reactionary forces; the Acceleration of space wrt mass also creates $F = ma$ reactionary forces.[5]

---

[5]Einstein's Doctrine of Relative Acceleration: When working out the Theory of GR, Einstein reasoned that the force experienced by a body accelerated wrt to the universe would be indistinguishable from that experienced by the same body at rest in an accelerating universe.

The pressure created by the reactionary force of accelerating space acting upon M_B must equal the pressure created by the reactionary force of accelerating space acting upon the operative inertial plane σ_U of the universe.

The Descriptive In Summary

The universe does not act as a uniform density sphere - being potentially infinite in extent, space functions operationally as an infinite inertial plane, or more accurately, an ensemble of dynamically responsive infinite planes, reflecting in total the inertial modulus σ_U (the modulus of inertial space as the impediment of changing momentum). As an inertial property of space, σ_U it exists as a constellation of many planes each having infinitesimal surface densities. Because the reactionary force of infinite planes is independent of distance, each plane offers instantaneous response along a line of action perpendicular to the plane. There can be no net force or net energy in a self creating cosmology. For unidirectional acceleration of space or mass, only one orthogonal inertial plane is required to explain action and reaction. Where gravity is involved, the universe must respond in 3 dimensions, ergo, cosmological response to interference with isotropic expansion is 3-D. The mathematical construct, is physical only in the sense it produces a dynamic impedance crafted from real Hubble parameters. As functional operatives, infinite planes are the means by which the conservation laws of momentum and energy are expedited. Gravity, like other reactionary forces, is a real force.

Gravitation From Expansion

Derivations and Equations

The evidence for the pseudo force theory of gravity follows from its predictions. Beginning with the well feted laws of motion put forth by Isaac Newton, the force of the universe acting upon a body 'B' is equal to the force of the body 'B' acting upon the universe.

Applying Newton's 2nd Law, force can be expressed in terms of an acceleration and a mass: If M_U symbolizes the effective mass of the universe, then:

$$M_B(a_n) = M_U(a_2) \tag{P-1}$$

where M_B is the mass of 'B' and 'a_n' is the isotropic acceleration of space created by global expansion. The objective will be to determined the counter acceleration of the universe a_2 required to balance the equation. For an infinite universe, M_U is not meaningful, but the equation can be made definite if both sides are represented in terms of *"mass per unit area."*

For a universe homogeneous on the cosmological scale, the effective mass can be expressed as an average density taken from a spatial sample of sufficient size, stipulated herein as the Hubble sphere. Accordingly, when both sides of equation (P-1) are normalized to a mass factor having a common unit area of one square meter, then (P-1) becomes:

:
$$\frac{M_B a_n}{m^2} = \frac{M_U g_2}{m^2} \tag{P-2}$$

Where 'a_2' now has a different value ('g_2') determinable in terms of the ratio of surface densities. To simplify the numerical process,' M_B' will be considered a uniform density sphere of radius 'r' and area $4\pi r^2$. Distributing Hubble mass M_U over Hubble surface area $4\pi R^2$ creates a shell density $\sigma_U = (M_U/4\pi R^2)$. Distributing M_B over the area $4\pi r^2$ creates a shell density $\sigma_B = M_B/4\pi r^2$

It will be understood that force obtained by considering the Hubble from afar as a uniform density volume ρ_U does not lead to the same gravitational field strength as that for a shell having the same mass and size.[6] That the 2-sphere gravity field will be different from the 3-sphere construct, must be taken into account when transforming from one to the other. Transformation from 2-sphere to a flat plane, reduces binding energy to zero. But because expansion is bidirectional, the effect of the infinite flat operative is felt twice - once in each direction -- consequently the strength of the reaction is identical to the 2-sphere. Rewriting (P-2) in terms of pressures:

$$\frac{M_B a_n}{4\pi r^2} = \sigma_B a_n = \frac{M_U}{4\pi R^2} = \sigma_U g \tag{P-3}$$

From which

$$g = \frac{\sigma_B}{\sigma_U}\left[a_n\right] \tag{P-4}$$

---

[6]The estimated value of $R = 1.1 \times 10^{26}$ meters is smaller than the estimated value of the 3-D Hubble scale $R_H = (1.3 \times 10^{26}$ meters) based upon $H_o = 70$ by a factor of 5/6. This difference arises because of the difference in the gravitational binding energy that arises when w formulating the operative form of the Hubble mass as a 2-D surface density. as for the loss of gravitational binding energy when transforming from a 3-sphere to a 2-sphere volumetric adjusted for change to the estimated size of a sphere formed corresponds which corresponds surface area of 1.5×10^{53} kg.

From Newton's Law of gravity, the intensity of the field at a distance 'r' from center of a spherical mass M_B is:

$$g = M_B G / r^2 \qquad \text{(P-5)}$$

Thence, combining (P-5) and (P-4):

$$G = \frac{r^2 g}{M_B} = \frac{r^2 \sigma_B [a_n]}{M_B \sigma_U} = \frac{r^2 M_B [a_n]}{4\pi r^2 \sigma_U M_B} = \frac{[a_n]}{4\pi \sigma_U} \qquad \text{(P-6)}$$

We are now in a position to calculate G from Hubble parameters. Two factors need to be determined. The cosmological acceleration factor 'a_n' can be expressed in terms of the Hubble scale R and the rate of recession of the Hubble surface. More specifically, the change in the radial rate of expansion is obtained from so called deceleration parameter 'q' defined as:[7]

$$q = (-) \frac{\ddot{R}(R)}{(\dot{R})^2} \qquad \text{(P-7)}$$

Since it is now widely accepted that the expansion rate is increasing, 'q' will have a negative sign and a value of unity, i.e., $q = (-1)$, accordingly:[8]

$$a_n = \ddot{R} = \frac{\dot{R}^2}{R} = \frac{c^2}{R} \qquad \text{(P-8)}$$

Where R is the operative value of the spatial scale that comports with the transformation of volumetric density to a surface integral (3-sphere —> 2-sphere). Hence (P-6) can be written as:

$$G = \frac{c^2}{4\pi R \sigma_U} \qquad \text{(P-9)}$$

That G encodes the cosmological acceleration factor c^2/R follows from its dynamic dimensionality $(m^3/sec^2)/kg$. That it is inversely dependent upon the operative scale 'R' imposes a reciprocality upon the inertial aliquot of M in an expanding universe. Simply put, the energy of a fixed number of atoms increases proportionally with 'R,' consequently the MG product remains constant.

---

[7]Misnamed *"deceleration parameter"* at a time when cosmologists were convinced expansion would be slowed by gravitational effects. For an accelerating cosmos, $q = (-1)$

[8]It should not be surprising that G, having dimensionality meters^3/sec^2, encodes the cosmological acceleration factor c^2/R. That the format of (P-8) comports with the acceleration created by a rotational velocity 'c' at radius 'R' will prove to be of value in understanding the nature of expansion.

As more fully developed *infra,* the effect of the dependency of $\mathbf{G} \propto (\mathbf{1/R})$ and the effect of $\mathbf{M} \propto (\mathbf{R})$ is unnoticed because neither can be measured separately. That inertia $\mathbf{M}$ and $\mathbf{G}$ are scale dependent variables, is evidenced in several cosmological relationships yet to be explained. Of long standing interest is the ratio:

$$\frac{\mathbf{M_U G}}{\mathbf{R c^2}} = \mathbf{1} \qquad (P\text{-}10)$$

The enigma is, within the limits of experimental error, why Hubble Mass $\mathbf{M_U}$ multiplied by $\mathbf{G}$ equals Hubble scale $\mathbf{R}$ multiplied by light velocity squared? A coincidence of our present era, or a truth for the past as well as the future? If the latter, then the numerator must increase proportionally with $\mathbf{R}$. But any variation in $\mathbf{G}$ or cosmic mass $\mathbf{M_U}$ runs afoul of standard theory. The present development, being *ab initio* at odds with standard theory, frees us to proceed unrepentant in suggesting $\mathbf{M_U}$ and $\mathbf{G}$ reciprocally variant parameters that gain or lose value in such a way as satisfy both (P-9) and (P-10). Specifically:

$$\mathbf{M_U} \propto \mathbf{R^2}, \text{ and } \mathbf{G} \propto (\mathbf{1/R}) \qquad (P\text{-}10A)$$

all of which redirects attention to $\boldsymbol{\sigma_U}$ in the roll of a free space inertial modulus. Operatively, $\boldsymbol{\sigma_U}$ represents the amalgamation mass and space into a common functionality. The gravitational reactive force can thus be understood as an expansion engendering imbalance between $\boldsymbol{\sigma_U}$ (in its roll as the Machian inertial representative of the cosmos as a whole) and the inertial reaction of a local body to the isotopic acceleration field created by global expansion. That gravitational forces, like inertial forces, are proportional to the individual mass of the body, places a theory falsifying constrain upon $\boldsymbol{\sigma_U}$ - its place in formulation of 'g' fields cannot impact the numerical value of $\mathbf{G}$ least the equivalence between gravitational and inertial mass be lost. In words, $\boldsymbol{\sigma_U}$ must be a unity factor.

That the density of the infinite flat plane operative is approximately 1 kg/m² is incidentally satisfied by creating a surface density using Hubble mass and manifold as representative parameters.[9] More extensively as described infra, but briefly for present purposes, the infinite plane density is determined by what is required to create a gravitationally equivalent spherical shell of area density:

$$\sigma_U = \frac{\mathbf{M_U}}{\mathbf{4\pi R^2}} = \frac{\mathbf{1,5 \times 10^{53}}}{\mathbf{4\pi (1..1 \times 10^{26})^2}} \approx \frac{\mathbf{1\ kg}}{\mathbf{m^2}} \qquad (P\text{-}11)$$

---

[9]Bare Hubble mass is the mass that would be measured if the particles were spread sufficiently apart to avoid the creation of additional gravitational energy produced by the action of gravity acting upon gravity. The operative geometry of the metaphorical infinite plane is flat and consequently devoid of internal binding energy. Based upon Galactic density, bare Hubble mass is taken as 1.5×10^{53} kg. The effective scale (1.1×10^{26} meters) is based upon the Hubble scale (1.3×10^{26} meters), this being the adjusted size of a spherical shell that creates the same gravitational intensity as the 3-D Hubble sphere.

From (P-10)

$$G = \frac{c^2 R}{M_U} = \frac{c^2 R}{\rho_U V} = \left(\frac{c^2 R}{\rho_U} \times \frac{3}{4\pi R^3} \right) = \frac{3c^2}{4\pi R^2 \rho_U} = \frac{3H^2}{4\pi \rho_U}$$ (P-12)

The last expression will be recognized straightaway as Friedmann's espression for the density of a **q = -1** universe. To express **G** in terms of σ_U, we transform volumetric density to surface density: Since (P-11) derived from 3-D Hubble parameters adjusted for the difference in the gravitational energies between the 2-sphere and 3-sphere then (P-12) can be written in terms σ_U as:

$$G = \frac{c^2}{4\pi R \sigma_U}$$ (P-13)

The (P-9) derivation of **G** that began with Newton's 2nd law (P-1) and the (P-13) derivation of **G** which evolved from the long puzzled over relationship between volumetric acceleration and Hubble mass (P-10), are identical. When **G** is formulated as in (P-12), then:

$$G = \frac{Rc^2}{M_U} = \frac{c^2}{4\pi R \sigma_U}$$ (P-14)

Consequently per (P-10A).: $$M_U = 4\pi R^2 \sigma_U$$ (P-15)

In the imaginary tug-of-war between expansion and gravity, victory was historically believed to depend upon the amount of mass in the universe. Formulated, in terms of a density factor Ω., much effort was made to access the critical value ρ_c which made $\Omega = 1$. The idea of fine tuned balance between gravity and expansion is superseded by a derivation of the former in terms of the latter.[10] Since gravity derives from expansion, the mass of the universe, and consequently the omega factor Ω, no longer have a say in the fate of the universe. The idea of critical density should have been discarded after the discovery of exponential expansion, but the guardians of the standard model turned instead to a search for undiscovered energy forms. Omega will always appear near one, because the root cause of gravity is expansion. With that knowledge, comes the answer to the great cosmological profundity —> the ultimate fate of the universe. Without expansion, there is no gravity, and without gravity there is no final collapse. In the end, there is no end.

---

[10]Past studies of recessional velocities were interpreted as a cosmic state of critical density (in words, a diminishing expansion velocity destined to reached zero in infinite time). The consensus was both mathematically elegant, and emotionally appealing. The cosmos was seen as governed by the same principles as those used to calculate the escape velocity of projectiles launched from the earths surface. The exponentially decelerating model of the universe (First studied by Einstein and de Sitter, circa 1932) became the defacto standard. It is customary to express the density as a fraction of the density required for the critical condition with the parameter $\Omega = \rho/\rho_{critical}$ so that $\Omega = 1$ represents the condition of critical density. The logical paradigm that gravity follows from expansion, would seem to have escaped serious inquiry, the focus has been upon a cosmological conflict that never existed.

There remains, however, competition between expansion and gravity — the distance at which mass can capture matter. Specifically:

$$\frac{MGM^*}{r^2} > \frac{M^*c^2}{R} \qquad \text{(P-16)}$$

Substituting for G from (P-13):

$$\frac{Mc^2}{r^2 \, 4\pi R\sigma_U} > \frac{c^2}{R} \qquad \text{(P-17)}$$

Whence

$$M > 4\pi r^2 \sigma_U \text{ and therefore } r_c = \sqrt{M/4\pi\sigma_U} \qquad \text{(P-18)}$$

From (P-18), there is a critical distance r_c where gravity dominates over its Dominus.

From its theoretical inception in 1917 to its eventual discovery in 1998, the accelerating universe slumbered as a factual reality beneath the predominate view of gravity as a primary and independent property of mass induced curvature.[11] The unification of gravity and expansion is made complete with the inclusion of the $(q = -1)$ exponential expansion parameter c^2/R in the formulation of G [equations (P-6) and P-13)]. Long unresolved inconsistencies are disposed, puzzling congruencies are explained. Was it not obvious that every predictive theory of gravity required the insertion of an experimentally measured acceleration factor G? Could Einstein have been so chagrined by the discovery of expansion that he failed to see the significance of the cosmological constant, not as a volumetric acceleration to balance gravity, but as the instrument of its creation?[12]

---

[11]While Einstein's *"Equivalence Principle"* disposed of the need for a separate gravitational mass *per se*, he did so by imposing a new and unproven physical principle, namely that inert matter bends static space. There was no physics to support the proposition, moreover the idea of bent static space would soon have to be re conceptualized in the light of Hubble's discoveries. For lack of an alternative explanation of gravitational cause, curved space is still commonly taught and depicted pictorially as a weight supported by an elastic fabric. For many cosmologists, the analogy is considered metaphorical at best. There is no question that "time" is affected by a gravitational field, and since $\Delta s = c[\Delta t]$, there is strong reason to believe space is in someway influenced by matter.

[12] In late 1916, Einstein published his final draft of General Relativity, which included for the first time, an *ad hoc* cosmological constant Λ to prevent his static model of the universe from collapsing. To balance gravity, Λ would have to have the value $3H^2$. After the discovery of expansion, Einstein was remorse (in that by introducing the CC he had missed the opportunity to predict the expansion of the universe). He suggested Λ be done away with, since an expanding universe needs no static solution and consequently no gravity balancing factor. But it was soon realized that the expanding universe had a balancing problem of its own. After billions of years, it could not be determined whether the present rate of expansion was increasing or decreasing. Within the limit of the experimental techniques available throughout most of the 20th century, the recessional rate appeared to be constant. This lead to the idea of a finely tuned relationship between expansion and gravity, which surprisingly still prevails. Willem de Sitter, only months after its publication, discovered an exponentially expanding solution based upon the CC. In its initial formulation, space was devoid of both pressure and matter. Howard Robertson later showed the solution would be valid for a low density cosmos comprised of sprinkled clumps of matter (our universe). Einstein could have reinterpreted the CC as the dynamic balancing factor between Gravity and Expansion. Or as the essence of gravity, since to balance gravity Λ had to have the value $4\pi G\rho_U$, thence: $G = \Lambda/4\pi\rho_U = 3H^2/4\pi\rho_U = c^2/4\pi R\sigma_U$

The failure of gravity to exert a retarding influence upon space, the implication of the cosmological constant Λ in the formulation of spatial expansion, and the encoding of $\Lambda = 3H^2R$ in the derivation of **G**, were all previously hidden truths unveiled by one unexpected discovery.

An operative infinite plane, can in theory, only be accommodated by an infinite universe. To maintain zero energy, the cosmological constant Λ creates negative energy reactive fields in equilibrium with maturating inertial energy sired by expanding negative pressure. The cosmological constant defines empty space as an operative. In truth, it is one of two known characteristic of the vacuum, providing as it does the emblement for self creation.

The impact of Friedmann's early work in combination with Λ, can alas, be fully appreciated. Originally distilled from Einstein's final version of General Relativity (and subsequently rediscovered independently by the Belgian Priest, Georges Lemaitre), the same equations were later derived in 1934 by William McCrea and Edward Milne using only Λ and the physics applicable to of a free falling body in a gravitational field.[13] What is remarkable, when positive mass energy ρ_U equals negative pressure energy $-3P/c^2$, the equations reduce solely to the operative effect of Λ.[14]

$$\ddot{R} = -\frac{4\pi G}{3}\left(\rho_U + \frac{3P}{c^2}\right)R + \frac{\Lambda R}{3} \dotfill (P-19A)$$

$$\dot{R}^2 = \frac{8\pi G\rho_U R^2}{3} + \frac{\Lambda R^2}{3} - k \dotfill (P-19B)$$

[13]In the years immediately following the publication of GR, only the ablest of theoreticians were able to find solutions applicable to the universe as a whole. But in 1934, two Cosmologists in Britain, Edward Milne and William McCrea, showed that the equations controlling the dynamics of the universe could be derived directly from simple Newtonian theory. This surprising development created a puzzle that is yet to be resolved. Why should two such different formalisms lead to the same expressions? In one sense, it can be reasoned that a theory is better evidenced if it can be arrived at by independent means. Arguendo, such occurrences would be inevitable in a zero energy universe. That zero energy is an initial condition as well as an ongoing criteria, evokes a deeper profundity? Perhaps what appears to be a holistic cosmos is nothing more than the evolutionary manifest of nothing.?

[14]In an exponentially expanding universe, the Hubble term:

$$H = \frac{\dot{R}}{R}$$

is constant because the rate of change of the scale is proportional to the scale. In a flat universe, the curvature $K = k/R^2$

Assuming spatial density ρ_U and pressure P is the same everywhere. When Λ is included, the equation of motion of a radially moving particle of unit mass at the surface of a sphere of radius 'R' is:

$$\frac{dv}{dt} = \ddot{R} = -\frac{4\pi G r_U[R]}{3} + \frac{\Lambda R}{3} \tag{P-20}$$

Integrating (P-20) gives:

$$\dot{R}^2 = \frac{8\pi G \rho_U R^2}{3} + \frac{\Lambda R^2}{3} - k \tag{P-21}$$

In the zero energy expanding universe, energy due to pressure is a critical factor in explaining the evolution of the cosmos. Accordingly, when (P-20) is modified to include pressure energy:

$$\ddot{R} = -\frac{4\pi G}{3}\left(\rho_U + \frac{3P}{c^2}\right)R + \frac{\Lambda R}{3} \tag{P-22}$$

which is the Friedmann equation (P-19A). The Pressure $3P/c^2$ corresponds to the pressure energy density, expressed in units of mass. P like ρ_U, may exist in different forms, some of which are cancelling in their gravitational effect. For example, radiation pressure is positive, gravitational pressure is negative, expansion of negative pressure creates positive energy, whereas spatial expansion itself, increases the volume of negative energy fields.[15]

In a radiation dominated era, positive pressure plays a significant role. If all cosmic energy existed as radiation, ρ_U would be zero in (P-22). To preserve the mandate of zero energy at all epoch's, the hypothesized beginning state of radiation only would necessarily be accompanied by an equal negative energy pressure (that required by expanding space to create the radiation).[16] Rephrased, the positive energy contained in a spatial volume of radiation would equal the negative pressure energy that went into creating the radiation contained in the volume.[17]

---

[15]Following his 1934 work with Edward Milne reconstructing Friedman's work from Newtonian principles, McCrea proposed a new model of the universe based upon spatial tension. The surprising mathematical consequence of a negative pressure cosmos is that positive energy is created during expansion. The Λ parameter suddenly finds intelligible physical coherence. When tension equals 1/3 the energy density ρ_U ceases to have any effect -- the dynamics of the universe are determined solely by the cosmological constant.

[16]A vessel of volume V having perfect reflecting walls for containing a radiation mass density 'ρ,' will weight $2\rho V$ rather than ρV. The common explanation is that the internal radiation has exerted stress upon the walls. These stresses are a form of energy that equals $3PV$.

[17]In the usual case, radiation pressure confined to a container exerts additional positive pressure on the walls of the containing vessel equal to the radiation energy contained in the volume. In expanding space, there are no walls as such, nonetheless from a cosmodynamic perspective, expanding space creates negative pressure. From the development of (P-3), the pressure should be relatable to the product of the acceleration times the inertial

reactance of the cosmos, i.e., $\qquad -P = (a_n)\sigma_U \qquad$ (P-24)

There are of course, no real container defined volumes in space, nonetheless radiation behaves gravitationally as a mass $2\rho_U V$ per (P-21). When a volume contains only radiation, half of the gravitational mass $2\rho_U V$ is in the form of expanding negative energy (that which created the spatial tension stress that caused the formation of the radiation particle), ergo, energy density is zero.

The criteria for zero energy is simple and efficacious. Net energy is zero because $\rho_U = -3P/c^2$. To check this proposition, the pressure energy $-3P/c^2$ should equal positive energy ρ_U. From (P-24):

$$\mathbf{E_U} = -\frac{3P}{c^2}[V] = 3[a_n]\frac{\sigma_U}{c^2}\frac{4\pi R^3}{3} = 3\frac{c^2}{R}\frac{\sigma_U}{c^2}\frac{4\pi R^3}{3} = 4\pi R^2 \sigma_U \qquad (P-23)$$

Comparison of (P-23) with (P15) shows that the energy $\mathbf{E_U}$ obtained by multiplying cosmic negative pressure energy density $-3P/c^2$ by the volume of the Hubble sphere $\mathbf{V}$ equates to the positive mass energy $\mathbf{M_U}$ obtained from the derivation of $\mathbf{G}$ using Hubble parameters.[18] From Friedmann's equation, critical density is:

$$\rho_c = \frac{3H^2}{4\pi G} = \frac{3c^2}{4\pi R^2 \left[\dfrac{c^2}{4\pi R \sigma_U}\right]} = \frac{3}{R}[\sigma_u] \qquad (P-25)$$

Which from (P-15) or (P- 23), corresponds to:

$$\rho_U = \frac{M_u}{V_U} = \frac{4\pi R^2 \sigma_U}{\dfrac{4\pi R^3}{3}} = \frac{3\sigma_U}{R} \qquad (P-26)$$

That (P-25) and (P-26) are identities, resolves the omega enigma. Derivation of the critical density parameter $\mathbf{\Omega} = 1$ by mathematical means, is consistent with our initial propositions, that of the dependence of gravity upon the rate of expansion.

---

[18]Comparison of (P-25) with the derivation of vacuum density based upon the ratio of Planck energy to Planck volume leads to:

$$\rho_{vacuum} = E_p/l_p^3 \qquad (P-27)$$

where $l_p = [(h/2\pi)(G/c^3)]^{\frac{1}{2}} = 1.6 \times 10^{-35}$ meters (The Planck length) and $E_p = [(h/2\pi)(c^5/G)]^{\frac{1}{2}} = 2 \times 10^9$ Joules, is the Plank energy. That this exercise leads to vacuum density 120 orders of magnitude larger than that obtained using Friedmann's equations, would seem to be more good cause for suspecting Planck units to be nothing more than numerological curiosities, devoid of predictive power.

Omega (Ω) is unity, not because of fine tuning, but because gravity is expansion dependent inertial reaction.[19] With the recognition of gravity as the pseudo force accompaniment of spatial dilation, the path has come full circle, back to the cosmological constant. Formulated as $\Lambda R/3$, it prescribes the exponential ontogeny of the void. The zero energy universe has but one solution, i.e., when:

$$\rho_U = \frac{-3P}{c^2} \tag{P-28}$$

Then (P-22) reduces to:

$$\ddot{R} = \frac{\dot{R}^2}{R^2} = \frac{\Lambda}{3} = \frac{3H^2}{3} \tag{P-29}$$

Where Einstein's value $3H^2$ has been substituted for Λ.
Taking square root then:

$$\int \frac{dR}{R} = \int H(dt) \tag{P-30}$$

Whence:

$$R = e^{Ht} \tag{P-31}$$

As an initial condition, the zero energy state slumbers as an instability with the potential to self create. Once set in motion, the runaway condition is fueled by expansion of negative pressure space. Expansion of stressed space begets negative and positive energy equally. The exponential solution (P-31) speaks for itself, no dark energy needed. When asked to summarize GR in one sentence, Einstein replied:

"Time, space and gravitation have no separate existence from matter...physical objects are not in space. But these objects are spatially extended."

---

[19]Emergent fields are not new, but historically they have proved to be a dangerous proposition that provoked theologians. The preservation of his own life was likely the motivation for the guarded language and credit artfully bestowed upon the creator in this 17th century manuscript:

"...the action by which he (GOD) now sustains it is the same with that by which he originally created it; so that even although he had from the beginning given it no other form than chaos, provided only he had established certain laws of nature and had lent it his concurrence to enable it to act as it wont to do, it may be believed, without discredit to the miracle of creation, that in this way alone, things purely material might, in the course of time, have become such as we observe them at present; and their nature is much more easily conceived, when they are beheld coming in this manner gradually into existence, than when they are only considered as produced at once in a finished and perfect state."

Rene Descartes (1637)

Unification of gravity as *"expansion-induced"* reactionary force, reveals Λ to be the source of **G**. The equipoised state of positive and negative energy, materializes as exponentially expanding flat inertial space, metaphorically in the form of infinite planes which together proffer a composite density σ_U, but physiologically as a nearly vacuous 3-D volume. In this roundabout regenerative way, self perpetuating cosmology emerges as intelligible.

While the positive feedback loop for continued cosmic expansion is bolstered by established physics principles, genesis ex nihilo [based upon a mathematical relationship] that assures the continuity of continued existence once started, also requires a retro that brings about the change that kicks off the process. Now called 'time' or *temporal progression,* then *'expanding space'* once started, requires for genesis impetus, the proposition that physical reality can be founded upon an abstract metaphysical potentiality[20] Justification for the inevitable universe turns upon the question of whether the *"now"* rate of expansion $\{c^2/R\}$ as an abstract mathematical presence, is sufficient to initiate a physical beginning from nothing.

If so, was their a creation state of infinite acceleration c^2/r when $r \rightarrow 0$ for an infinitesimally short duration? Credible evidence and well reasoned theoretical predictions support the prevailing *"bang"* view of the universe. That the universe once existed as a hot dense radiation phase is of little doubt, but the manner of its evolution into and out of such a condition, is yet much debated. In a lilliputian universe having a cosmic scale r (small in comparison with the present **R**), the inverse dependence of the expansion algorithm upon scale translates to infinite start-up stress for an infinitesimally short period of time. By any theory of creation stemming from expanding negative pressure, the vacuum required for the production of radiative particles, is satisfied by a sufficiently small spatial scale – in such a case, creation *Ex nihilo* may be imminent?[21] For a zero energy universe, both positive and negative energy would thereafter increase as R^2

In the zero energy universe, inertia is enhanced because the energy in the spatial expansion field is increased. Negative energy accumulated by spatial expansion stress is reflexed to the particle as enhanced inertial mass. Contrary to traditional theory, mass may be nothing more than a manifest of the negative energy stress field (the extension of a particles '**g**' field throughout the volume of the universe) which is powered by continuous spatial expansion.

Standard theory tries to make sense out of a universe built from fixed constants. In reality, very few ratios have been experimental confirmed to be immune from change. While the space/time ratio '**c**' and the dimensionless angular momentum ratio 'α,' appear to be temporarily invariant, these are ratios (no single quality or property alone (e.g., mass) is invariant). Indeed, the implication of the cosmic relationship $M_U G/R = c^2$ is that either **G** or M_U (or both) must change as the scale factor increases.

---

[20] Faith defined by American Humorist, Mark Twain: *"Is believing in something you know just ain't so."*

[21] Negative pressure diminishes as volume increases. The creation of radiative particles diminishes as negative pressure palliates. Production of radiation and proto particles cannot be sustained beyond a certain negative pressure level, and stops without intervention when spatial stress energy drops below a critical value. As space dilates, photon energy is lost, proto particles cool and combine to form matter and their accompany'**g**' fields. To illustrate, assuming σ_U is a kg/meter^2 constant of the expanding universe, then when **R** equals one meter, the intensity of the vacuum pressure is $\{-(3 \times 10^6 \text{ meters/sec})^2 \times (1 \text{ kg/meter}^2)\} = 9 \times 10^{12} \text{ ntn/meter}^2$

Abrupt beginnings present a problem for the zero energy theory of the universe. An alternative to a temporal genesis (some 14 billion years past), is suggested by the present rate of expansion c^2/R. The zero energy universe can be made consistent with a past eternal evolutionary process, subject to the interpretation the relativistic relationship between space and time, i.e.,

$$ds = [c]dt \qquad\qquad (P\text{-}32A)$$

Applied to universe, the claim can be made that the rate of change of time depends upon the rate of change of space. Taken on its face, it cannot be determined which is the independent variable. The rate of passage of time depends upon the size of the universe? And if the rate of passage of time depends upon the rate of spatial expansion, then our perception of time is reduced to our perception of the consequences of spatial change. The integral of $ds = r = ct$. The initial condition (zero energy) provides the intrinsic instability required for a beginning (but not a bang). When judged by the "now" rate of temporal flow as a constant, creation appears an event about 14 billion years past. From the perspective of time as space dependent, the universe is past eternal:

$$S = \int at(dt) = \int \frac{c^2}{r}\left[\frac{r}{c}\right]dt = \int c(dt) = R \qquad\qquad (P\text{-}32B)$$

If time proceeds at a constant rate, the Hubble constant will be constant in an accelerating universe.

$$H = \frac{\dot{R}}{R} = K$$

So (P-32B) can be written

$$1/H = r/c = K$$

And therefore

$$S = \int at(dt) = \int \frac{c^2}{r}\left[\frac{r}{c}\right]dt = K\int \frac{c^2}{r}dt = \frac{K}{K}\int c\, dt = R$$

when viewed from the proposition that the flow of time is constant, the universe is about 14 billion years old. If the progression of time depends upon the scale of space, then the universe is past eternal. Those predisposed toward creation theory and big bang beginnings, avoid dilemma, and count time as a constant rate of change. For those of an Eddington persuasion, measuring time as space dependent will satisfy their bent.[22] This is a curious formalism for an eternal universe with the countenance of a beginning. Start up is less abrupt, but it is more difficult to explain a hot dense phase of pre-particle production because pressure and stress do not change rapidly.

---

[22]George Lemaitre was chastised for his interpretation of the expanding universe as the creation of a deity, whereas Arthur Eddington was repelled by the genesis view of universe. The two men worshiped in different cosmological temples. Anti-bangers like Eddington, put their faith in the idea that evolution required infinite time to reach its present state (considered a necessary premise if the universe is to have a natural beginning). .

The essence of the inextricable tie between mass, space and time, is to be found in the manner by which the universe as a whole exerts its presence upon its parts. In what proceeded, the 3-D cosmos was refashioned by transformations to create two dimensional equivalents. In **Fig F**, the natural operative state of a flat space universe is as a plurality of infinite planes, each having thickness '**d**.' For a cube enclosing a mass **M** having a volume sufficiently large to represent average cosmic density, then if "**X**" is the length of a side, the volumetric density ρ_U is:

$$\rho_U = \frac{M}{X^3} \tag{P-33}$$

In **Fig 4**, the sample cube has been sliced into '**n**' imaginary square slabs parallel to each of the 3 faces. There are '**n**' slabs parallel to the x-y plane, '**n**' slabs parallel to the x-z plane and '**n**' slabs parallel to the y-z plane. Each slab functions collectively with all other slabs to oppose acceleration normal to its surface. Because the geometry is three dimensional, only 1/3 of the total mass **M** is available to oppose acceleration in any one direction:

$$\sigma_S = \frac{\rho_U X}{3(n)} = \frac{MX}{3X^3[n]} = \frac{M}{3X^2[n]} \tag{P-34}$$

For the universe as a whole, there will be '**n**' parallel slabs acting normal to the direction of any acceleration. All slabs act in concert with equal force to create a counter force opposing acceleration normal to their surface. The operative state of a flat space universe exists as parallel slabs which combine to exert a reactive impedance $\perp$ to their surface. The effective density of all slabs in the normal direction

$$\sigma_x = \sigma_y = \sigma_z = \frac{M[n]}{3X^2[n]} = \frac{M_U}{3X^2} \tag{P-35}$$

where M_U is the effective mass with the sample cube. Comparison with (P-15):

$$3X^2(\sigma_X) = 4\pi R^2(\sigma_U) \tag{P-36}$$

Whence for $(\sigma_X) = 1$

$$X = \left[\frac{4\pi R^2}{3}\right]^{1/2} = \frac{2R}{1}\left[\frac{\pi}{3}\right]^{1/2} \tag{P-37}$$

Fig 8

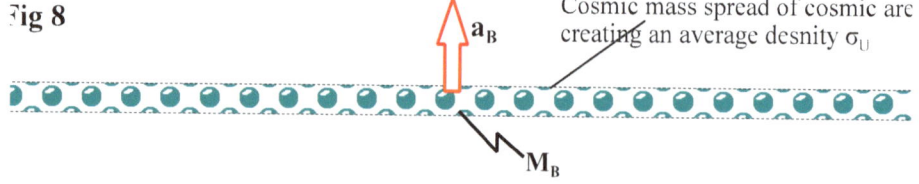

a_B

Cosmic mass spread of cosmic arc creating an average desnity σ_U

M_B

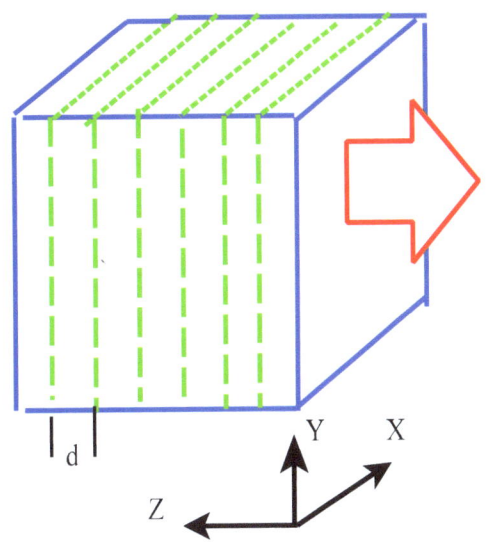

Fig 4. A cubical volume taken from a sample of Euclidean space realized as a number of slabs 'n' each having thickness 'd' = z/n and density ρ/yx, where "ρ" is the volumetric mass density (Total mass **M** of cube divided by the volume **xyz** .of the cube). If "ρ" is uniform throughout the cube, the area density σ of each slab is ρ(z/n)/(xy). A cubic sample commensurate with the extent of flat space will have **n** slabs and therefore since **x = y = z:**

$$\sigma = [\rho(z/n)/(xy)]n = \rho/z \qquad \text{(P-38)}$$

Fig 5 shows the expansion of a single slab stretched after four eons of expansion. No new matter is formed or lost so each triangular depiction of a section of a slab contains the same number of particles (whether electrons or galaxies). Each has increased in area and therefore the area density of the individual slabs has diminished inversely with the square of the change in length scale (x and y). But because the force created by infinite inertial planes is independent of distance from the plane, the effective inertial density of the universe based upon matter content, diminishes inversely with the square of the dimensional change rather than inversely with volume (Cosmic inertial reaction cannot be explained in terms of volumetric density). Because infinite slabs function only as area densities (asserting the same force throughout infinite flat space, it makes no difference which direction is chosen as the acceleration.

In **Fig 4** and **5**, acceleration is parallel to the **Z** axis, but the resulting inverse square dependence of inertia will be the same in any direction.

Fig 5

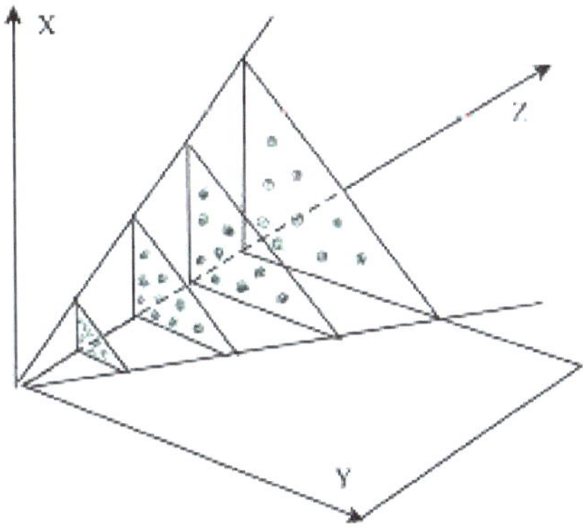

Gauss's law can be used to derive the gravitational field in cases where a direct application of Newton's Law is difficult (or in some cases impossible). The integral form:

$$\int_{\partial V} \mathbf{g} \cdot \mathbf{dA} = -4\pi\mathbf{GM}$$

(P-39)

For a Bouguer plate (an infinite flat plane of finite thickness), the gravitational field outside the plate is perpendicular thereto, with magnitude $2\pi\mathbf{G}$ times the mass per unit area, independent of the distance to the plate [There is no force until acceleration factor G is introduced]. In the absence of binding energy, a combination of two equal density parallel infinite slabs, produces no gravity in the space separating the planes, ergo, the gravitational affect of a plurality of parallel slabs (**Fig 4**) upon each other is zero (no adjustment for the effect of gravity acting upon gravity is required, the total mass per unit area $\boldsymbol{\sigma}_U$ is considered to be the result of bare mass)

Fig 6

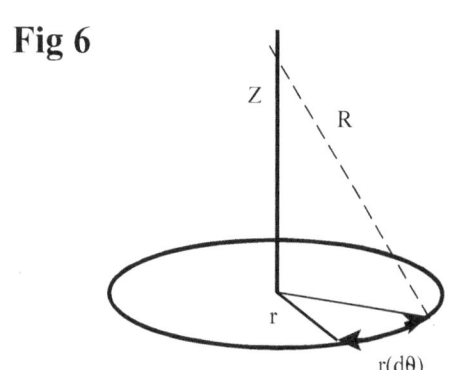

Fig 7

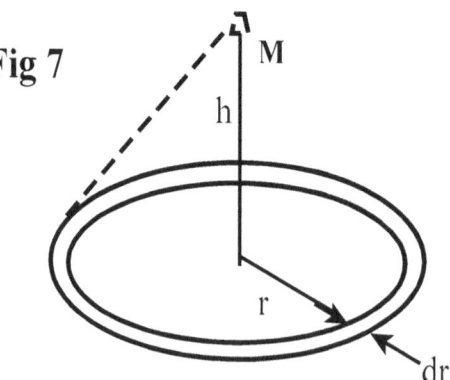

Fig 6 illustrates the geometry for finding the infinite flat plate potential at a point. We consider a disk and let the radius become infinite:

The gravitation potential for a disk of radius '**a**'

$$\Phi = -\mathbf{G}\int_S \frac{\sigma(\mathbf{ds})}{\mathbf{R}} = 2\sigma\pi\mathbf{G}\int_0^a \frac{\mathbf{r(dr)}}{(\mathbf{z}^2 + \mathbf{r}^2)^{1/2}}$$

When $\mathbf{r} \gg \mathbf{z}$, potential will not depend upon '**r**'

$$\Phi = 2\sigma\pi\mathbf{G}\mathbf{z}$$

(P-41)

Force lines will be perpendicular to the disk and the imaginary Gaussian surface S is pillbox (a short cylinder whose flat faces of area "A" are parallel to the plane. For the force F, at any height 'h' above the plane, we use Fig 7 and calculate the gravity acting upon a mass M by integrating the force felt by concentric rings from a radius zero to infinity. Expressed mathematically,

$$\mathbf{F/M} = 2\pi\mathbf{G}\sigma\mathbf{h}\int_0^\infty \frac{\mathbf{r}}{(\mathbf{h}^2 + \mathbf{r}^2)^{3/2}}\,\mathbf{dr} = 2\pi\mathbf{G}\sigma$$

So again, the force does not depend upon the distance '**h**' from the plane. The only factors that matter are **G** and '**σ**'

Inertia and gravitation are accounted for within the auspicious of a single field, σ_U. Initially constituted as (P-11) from the ratio of Hubble mass to Hubble area, now seen (P-38) apropos of flat space geometry. In the context thereof, σ_U abstrusely represents the reactance of the universe. That cosmic mass can be normalized in terms of spatial extent, parlays other analytical rewards, including the catalysing of **G** from **Λ**, the explication of Newton's second Law in terms thereof, and the accurate prediction of expansion engendered '**g**' fields emerged therefrom. In the infinite slab model of the universe, the effect of cosmic mass everywhere is made manifest as the inertial opposition to acceleration anywhere. Alas, Mach's principle is vindicated. Because σ_U is a unity operative [one kg/meter^2], it is fulfills the mandate that inertia is always proportional to mass.

With the recognition that flat space reactionary forces are explainable in terms of a large number of parallel slabs acting in concert, we arrive at **Fig 8** as the amalgamated space-mass structure of the universe. As an existing form, there is no need for further mathematical transformation (no volume integrals to surface integrals are required per the discussion of **Fig F** to explain cosmic counter action). Euclidean space exists in the form of a volume when viewed with respect to an acceleration in one direction and as an infinite plane (or a plurality of parallel infinite planes when viewed in a direction perpendicular to the acceleration). While at first glance bewildering, this will be understood as the logical consequence of inertial space.

On the large scale, the universe is smooth, on the small scale it is lumpy. To formulate the cosmos as a unified whole, space and mass must be fully homogenized. Lumps of matter dead ahead (until contacted) do not exert a retarding influence. Matter in the opposite direction exerts a gravitational retarding influence upon the acceleration, but this diminishes inversely with the square of the distance. To apply the inertial resistance of the cosmos to an accelerating body '**B**' anywhere in the universe, the force issued by the cosmos is independent of the location of '**B**.' A universe comprised of thin planes will act orthogonally to collectively oppose acceleration equally at any location. Space and lumpy mass amalgamate uniformly within the concept of the infinite plane.

As shown in Fig 8, the effect of other matter (M_0) along a line of action collinear with the accelerating motion of a body can only exert its presence by contact or gravity. In the void far removed from massive objects, the retarding force exerted by directly bumping into the few particles that make up the average density of the space, is insignificant. Opposition to acceleration, however, is always present indirectly in the mass of the universe as a whole. The action of global mass in the form of the infinite plane (or set of infinite planes) opposes acceleration of any of its members (e.g., M_B). All masses are part of the composite, and opposed by the density σ_U that it helps to create

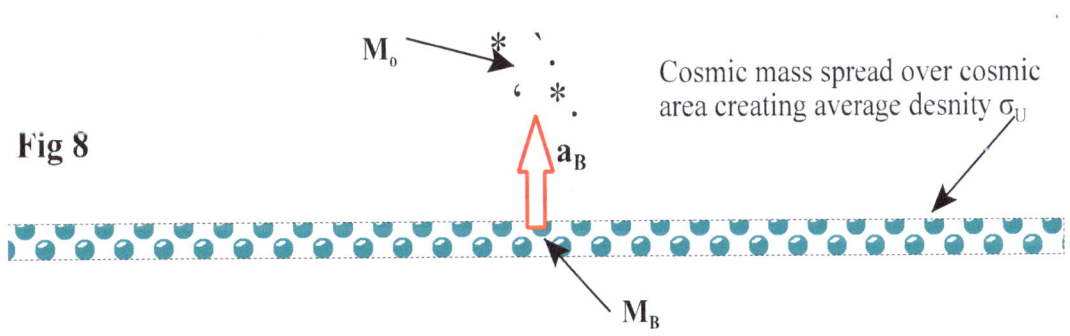

M$_0$

Cosmic mass spread over cosmic area creating average desnity σ_U

Fig 8

a$_B$

M$_B$

Elaboration on how Cosmic mass M_U opposes the accelerations of its constituency, is in order. Einstein's Principle of relative acceleration (footnote 5) offers analogy. Adverting again to **Fig 8**, the idea being that the force upon M_B will be no different if M_B is fixed while the remainder of the universe is accelerated (opposite to the red arrow). In this switch about of roles, the universe functions as an area density. Multiplied by acceleration, σ_U equals pressure (i.e., momentum flow). Required then, is to express M_B as an area density and equate the pressures in terms of their corresponding momentum flows. Specifically, if 'a_1' is the acceleration of mass M_B (idealized for tutorial purposes to have a uniform cross sectional area A_B orthogonal to the acceleration), the inertial reaction pressure $P_B = a_1 M_B / A_B$ where M_B is normalized in the same unit of area as cosmic mass (i.e., A_B is one square meter). Hence:

$$P_U = P_B \quad \therefore$$
$$a_2 \sigma_U = a_1 \sigma_B$$

(P- 43)

where (P-43) was previously derived in form (P-2) to express the gravitation force 'g' in terms of σ_U and the cosmological acceleration factor (where $a_1 = a_n = c^2/R$). Where mass M_B is accelerated wrt the universe, the integral of the spatial pressure over the surface equals the reactionary force. The pressure flux, of course, is not measurable as a surface phenomena as it acts directly upon the atoms and subatomic particles (wherever located within the interior) that comprise M_B (which as shown in **Fig 8**, is embedded in the universe as part of an infinite plane).

From Newton's 2nd Law, the acceleration of M_B wrt the universe requires no secondary acceleration of the universe [a_2] wrt to M_B (i.e., there is only one acceleration involved).[23] However, acceleration can also be expressed as **Force/kg**, so while pressure can be expressed as:

$$-\frac{\textbf{Force}}{\textbf{meter}^2} = \frac{-a_1 \sigma_B}{1}$$

(P-44)

When $\sigma_B \neq \sigma_U$, there exists a pseudo acceleration factor a_2 having dimensionality "Force/kg."

$$a_2 = \frac{\textbf{Force}}{\textbf{Mass}} = a_1 \frac{\sigma_B}{\sigma_U}$$

(P-45)

Equation (P-45) can only be true if σ_U equals (one kg/meter^2). Newton's 2nd law, by implication, then requires the operative scalar density inertial modulus σ_U to have a value one kg/meter^2. The symbolic spatial acceleration factor 'a_2' need not exist as a reality in the sense of spatial motion wrt mass? The credibility for the argument that Newton's second law is explainable in form as (P-45), is that space mimic's motion dimensionally. In the guise of asserting a counter force equal to the primary *acceleration-mass* product, 'σ_U' provides instantaneous local coupling to the cosmos as a whole.

---

[23]This assumes Newton's second law is dimensionally symmetrical - Taken literally, the force created by spatial acceleration relative to mass would need to be equal to the reactionary force created when mass is accelerated relative to empty space.

To complete the picture, **Fig 9** illustrates the effect of regarding the universe in its natural state as a unlimited volume acting as a continuum of parallel planes having unspecified thickness. The # of planes is immaterial, it is only the area density that is significant.

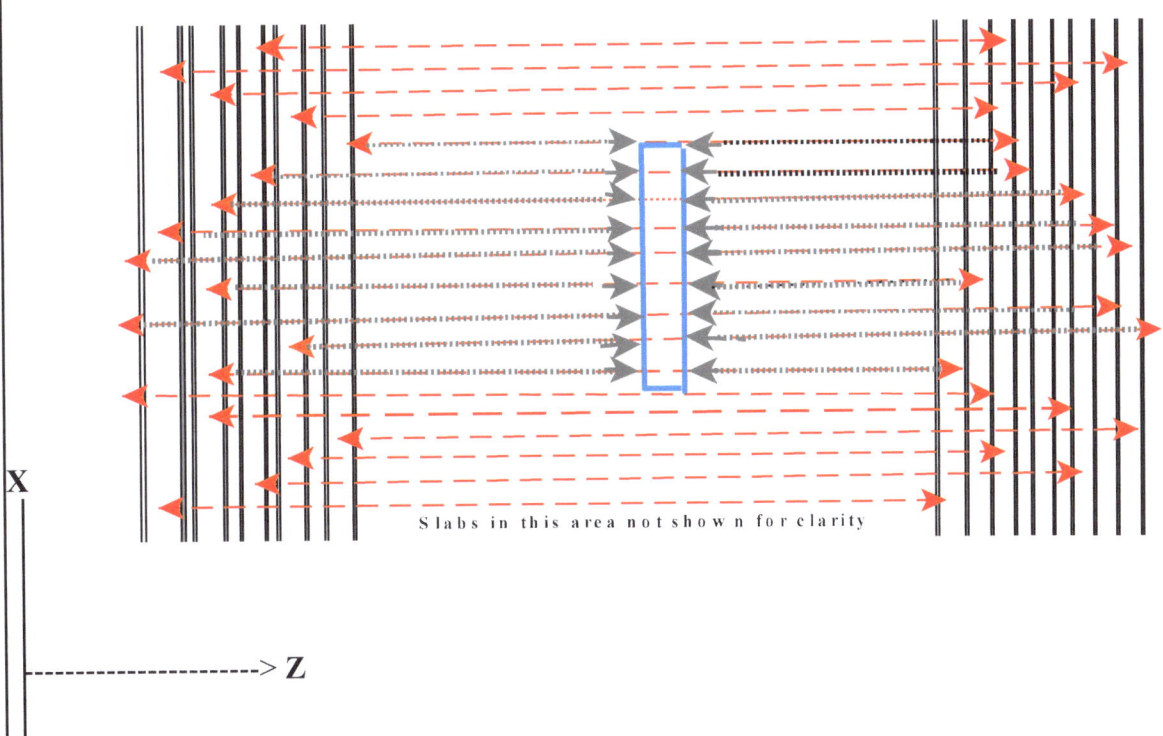

Slabs in this area not shown for clarity

X

-----------------------------> Z

Fig 9 is a cross section taken through the X-Z plane of a flat space universe. During expansion, the cosmic density per unit area remains constant. While additional planes are added by the growing size of the **Z** dimension, the **X-Y** area is increasing as the square. This condition would cause area density diminishes inverse square. However, new slabs are being added in the **Z** direction, and they are born in the environment of expanding negative pressure which creates positive energy (all new space is the result of expanding negative pressure). The net result is that the cosmic volumetric density (ρ_U) diminishes as (1/scale), total positive energy increases as (scale squared) and the area density σ_U by which gravity is determined, remains constant during expansion.

To calculate the 'g' field of a spherically uniform body having the same mass and size of the earth, it is convenient to imagine the earth's mass uniformly spread over a disk (blue) having the same area as the earth. By this ruse, the earths surface A_E defines the area of a cookie cutter sample of parallel flux lines emanating '$\perp$'to the cosmic slab stack). The flux pressure is therefore $[M_E/A_E]a_n$ and the cosmic counter force is $[M_U/A_U]g$ where the cosmic ratio was previously determined in terms of Hubble parameters. In general, for a body 'B' of mass M_B and surface area A_B, an interesting dimensional relationship emerges from the ratios:

$$\frac{a_n}{\sigma_U} = \frac{g}{\sigma_B} \tag{P-46}$$

-29-

Red dotted lines indicate acceleration lines which are shown as terminating at various slabs, illustrated by thin rectangles. In the center, the earth (blue) is shown edge-on as it would appear if its matter content were rearranged as a uniform flat disk in the X-Y plane. The earths inertial mass is an impediment to the free expansion of space between the slabs, shown by gray reaction arrows.

All the slabs in the universe are coupled to the cookie cutter area defined by the disk representation of the earth. Expansion lines are parallel and normal to the slabs for the situation where the inertial objects can be metaphorically depicted as a uniform flat surface. The dimensional parameters for the earth:

Earths radius = 6.37×10^6 meters
Surface area = $4\pi r^2 = 12.54(41 \times 10^{12})$ meters^2 = 510×10^{12} meters^2
Earth's mass = 5.98×10^{24} kg

The (P-46) ratios have the same dimensionality as **G**. Per (P-6), the left side ratio equals **4πG**. Ergo, the reactionary '**g**' field of a spherically symmetrical mass = **4πGσ$_B$**. In essence, modeling the universe as a stack of infinite planes reduces to Newton's Law of gravity. But Newton's Law does is not explain gravity, just as General Relativity does not explain gravity. Nor does Newton's 2nd Law offer a theory for inertial reaction. The slab stack infinite plane interpretation of flat space explains instantiated inertial reaction in terms of Hubble parameters. Does it do the same for gravity? If gravity depends upon inertia, then gravity is likewise instantaneous. In opposing acceleration, the cosmos does not act as a volume density, but as a stack of ∞ planes. There is of course, no real separation of the universe into slabs - the slab stack is an artifice -- the operative density of the universe is simply the result of adding up all the mass and dividing by the effective area.. Modeled as an operative stack of infinite planes that combine to present a unified ubiquitous instantiated reaction, the universe becomes intelligible.

Thus for the earth, using the modified scale R = 1.08×10^{26} meters to account for the loss of gravitational binding energy when transforming from volume to an infinite plane:

$$g = a_1 \frac{\sigma_E}{\sigma_U} = \frac{c^2}{R} \times \frac{\frac{5.98 \times 10^{24} \, kg}{510 \times 10^{12} \, meters^2}}{\frac{1kg}{meter^2}}$$

$$= \frac{9 \times 10^{16} \, (meter^2/sec^2)}{1.08 \times 10^{26} \, \textbf{meters}} \times \frac{0.0117 \times 10^{12} \, (\textbf{kg/meter}^2)}{1 \, \textbf{kg/meter}^2} = \frac{9.75 \, \textbf{meters}}{\textbf{sec}^2}$$

(P-47)

THE ORIGIN OF INERTIA

The universe acts as a infinite 2-D plane rather than a finite 3-D volume

We visually observe the universe as a three dimensional Hubble sphere. We experience the universe as a two dimensional Plane. Reactionary forces, including gravity, are the proportionality product of acceleration and mass, but strangely, inertial forces also dependent upon the mass of the universe. Indeed, it is the resistance of the void in opposing acceleration that reveals the inertial property of space as a dynamic modulus. For a body at rest in an expanding mass endowed space, gravity emerges as a pseudo force. On the cosmic scale, the universe is geometrically flat, ergo, the deceleration parameter 'q' is negative and equal to **-1**. The expansion rate expressed in terms of Hubble parameters is: .

$$\ddot{\mathbf{R}} = \frac{\dot{\mathbf{R}}^2}{\mathbf{R}} = \frac{\mathbf{c}^2}{\mathbf{R}} \qquad (1.1)$$

The concept of *"inertial space"* as a cosmological impedance reconfigures cosmic density in terms of area rather volume. More specifically, the operative geometry of the universe exists in unified form as space-mass amalgamation, configured not as volumetric density ρ_U, but rather as an area density σ_U comprised of 'n' infinite slabs which equally share the cosmic mass M_U. The slab model of flat space, requires no structural change – all inertial bodies remain in place, nothing is transformed, nothing is displaced. There are, however, two conceptual reforms to be absorbed:

1) That of how the universe organically marshals mass to oppose acceleration, and,

2) Why gravity fields emerge as inertial opposition to cosmological acceleration.
.

Accelerated masses in empty space experience reactive forces (one ntn/kg)/(meter/sec^2). How is this possible within the auspices of a density function in the range of **3 x10^{-26} kg/meter^3**? In lieu of reckoning cosmic mass uniformly distributed throughout the Hubble sphere, license is taken to reconstruct the cosmos piecemeal as a stack of cubes each having a volume **d^3** equal one cubic meter. Shown in **Fig 1** a single cube 'A$_1$' having density **10^{-26} kg/m^3** will exert a force upon a spherical body M_B displaced a distance 'd' from the cube center, of approximately G(M)(10^{-26})/d^2. Adding more cubes (**B$_1$, C$_1$, D$_1$ ...N**) directly in line with the first, per **Fig 2**, increases force, but because 'g' fields fall off inverse square with distance, the added force is small (no matter how many cubes are added, total force **F$_T$** is less than twice that produced by A$_1$ alone [**F$_T$ < [2G x10^{-26}M$_B$/d^2**].[24] Suppose instead, cubes are added vertically rather than horizontally. This creates a column of blocks above and below **A$_1$** as shown in **Fig 3**. Each new cube adds a small force at a less favorable angle, and because it is also further away from **M**, the effort appears to be even less productive than adding cubes horizontally -- that is, until the number of cubes 'n' --> ∞. When this occurs, the lines of force associated with the vertical column are perpendicular to the column.

---

[24]Known as the Basel Problem, originally proposed by Mengoli in 1644 and solved by Euler in 1734:

$$\sum_{n=1}^{n=\inf inity} \frac{1}{n^2} = \frac{1}{1^2} + \frac{2}{2^2} + \ldots \ldots \frac{1}{n^2} = \frac{\pi^2}{6}$$

The strength of 'g' field from gauss's law:

$$\int_S \mathbf{g} \cdot \mathbf{n} \ \mathbf{dA} = -4\pi\mathbf{Gm} \tag{1.2}$$

For an infinite column, the density of the force is determined using a cylindrical Gaussian enclosure of radius 'd' and length **L** placed to encompasses the cubical column concentric therewith. The column has a linear mass density $\lambda = 10^{-26}$ kg/meter, the flux is normal to the surface of the cubes, consequently, non escapes through the ends of the cylinder. Integration over the area of the cylinder gives:

$$-g(2\pi d)L = -4\pi Gm \tag{1.3}$$

Since **M** is the total mass enclosed by the surface **S**, then for a segment of length **L**, and density λ:

$$\mathbf{g} = \mathbf{2G\lambda/d} \tag{1.4}$$

The 'g' field falls off inversely with distance rather than inverse squared (per **Fig 2**).

Suppose while building vertically in the '**Z** direction, cubes are also added in the **Y** direction on each side of the vertical column to construct a one meter thick infinite slab parallel to the **YZ** plane. The slab has a mass to area density σ, the object then being to calculate the 'g' field at distance 'd' from the center of the slab. In this case the appropriate Gaussian surface is a pillbox -- a short cylinder whose flat end areas "A_P" are normal to the plane. The gravitational acceleration is normal to the outward normal unit vector, so the sides of the pillbox contribute nothing to the flux integral. The area integral in this case is $2\ A_P$ (only the two outward end areas contribute to the flux). Hence:

$$-\mathbf{g}\int_s \mathbf{dA} = -4\pi\mathbf{Gm} = -\mathbf{g}(2A_p) \tag{1.5}$$

The mass inclosed by **S** is a cookie cutter circle of area A_p and density σ, so it has mass σA_p. Therefore:

$$-\mathbf{g}(2A_p) = -4\pi\mathbf{G\sigma A_p} \tag{1.6}$$

Whence: $$\mathbf{g} = \mathbf{2\pi G\sigma} \tag{1.7}$$

For the infinite slab, gravitational acceleration is independent of the distance from the slab. By constructing a second, third and forth infinite slab parallel to the first, the force upon M_B is increased proportionally. Since each infinite slab exerts the same force upon M_B irrespective of the perpendicular distance from the wall, all slabs parallel to the **YZ** plane augment equally to increase the 'g' force upon M_B. By this contrivance, multiple **YZ** slabs can be functionally considered as a single infinite plane of density $2\pi Gn\sigma$ in contact with a mass (such as M_B) which is accelerated parallel to the '**X**' axis. Inertial reaction is instant, every slab makes its presence upon M_B continuously known (even though the slabs are physically distributed across the extent of the cosmos perpendicular to the direction of the acceleration.

While the operative principle as been illustrated using 'X' axis as the direction of acceleration, the density function is the same in any direction. Irrespective of the heading, an accelerating mass is impeded by all other mass energy amalgamated as an instantly local infinite plane σ_U orthogonal thereto. For an accelerating object, all other mass in the universe appears as an area density fashioned from ρ_U. That reactionary pressure can be tutorialized as a pseudo force created from a plurality of individual slabs, in reality there are no individual slabs. There is a single slab of thickness L and area L^2 having surface density M/L^2. The concept of the infinite plane as a dimension reducing operative can now be appreciated (the dimensionality of inertial space is operatively reduced from three to two). This is the physiology by which cosmic content is marshaled to oppose acceleration. Newton's 2nd law is now explainable in terms his 3rd Law.

$$F = M_U a_1 = M_B a_2 = \frac{M_U}{meter^2} g = \frac{M_B}{meter^2} a_2 \qquad (1.8)$$

where the 3rd ratio represents the reactionary force exerted by the universe upon an accelerating mass normalized in terms of a common area of one square meter and the 4th ratio is the force applied to accelerate a mass **B** also normalized in terms of the common area one square meter. It is known that the reactionary force per unit mass 'g' is one **ntn/kgm** for an acceleration 'a_2' equal to one meter/sec^2. Therefore :

$$\frac{M_U}{Meter^2} = \frac{a_2}{g} \; x \; \frac{M_B}{meter^2} = \frac{\dfrac{meter}{sec^2}}{\dfrac{ntn}{kg}} \; x \; \frac{kgm}{meter^2} = \sigma_U \qquad (1.9)$$

In order to comport with Newton's 2nd Law, the inertial modulus σ_U must equal one kg/meter^2.

To illustrate the action of the infinite plane universe from the perspective of an accelerated object, we consider a square flat sheet having mass M_B. As shown in Fig ___, when accelerated normal to its surface, the cookie cutter Gaussian pillbox is a square. When accelerated at an inclined angle, the sheet experiences the inertia of the universe as a rectangle, and when accelerated parallel to an edge, the universe is seen as an inertial line. As the cross sectional area of the pill box changes, the density normal to the direction of acceleration increases. The product of cross section area with cross sectional density is constant - when multiplied by $1/\sigma_U$, the force is always M_B x **a**

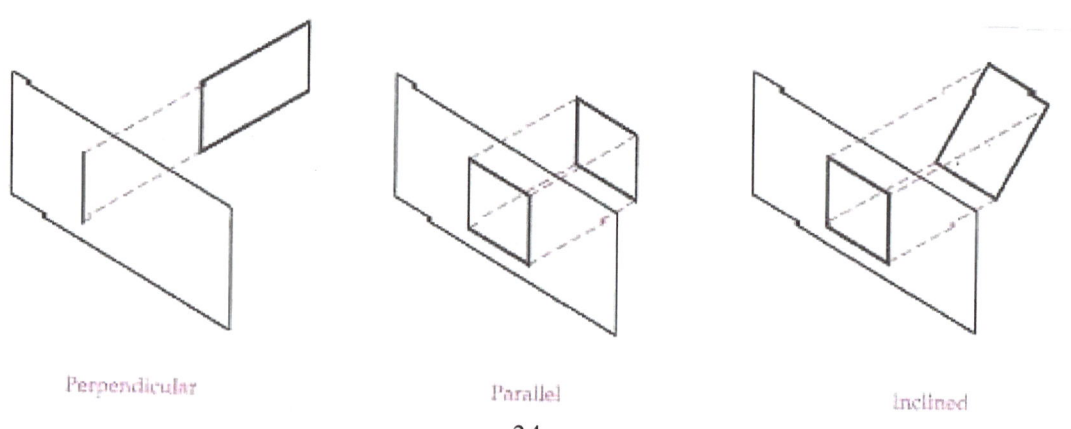

Perpendicular   Parallel   Inclined

-34-

In a similar vein, a uniform density 3-D body like that shown in Fig ___ will (for different directions of acceleration) punch out dissimilar shapes defined by their projection on the reactionary orthogonal plane σ_U that represents the universe in the direction of acceleration. Counter pressure exerted over the interface area between σ_U and cross sectional are of the body will be functionally dependent upon the depth of the body in the direction parallel to the acceleration (surface flux will increase in proportion to the depth profile measured parallel to the line of action). While natures minimum size for inertial quantization is not revealed by classical mechanics, the Planck scale looms as a candidate for the lower limit upon which Newton's 2nd law can operate as a discrete force. The composition of all forms of known matter are embraced within the classical formalism.

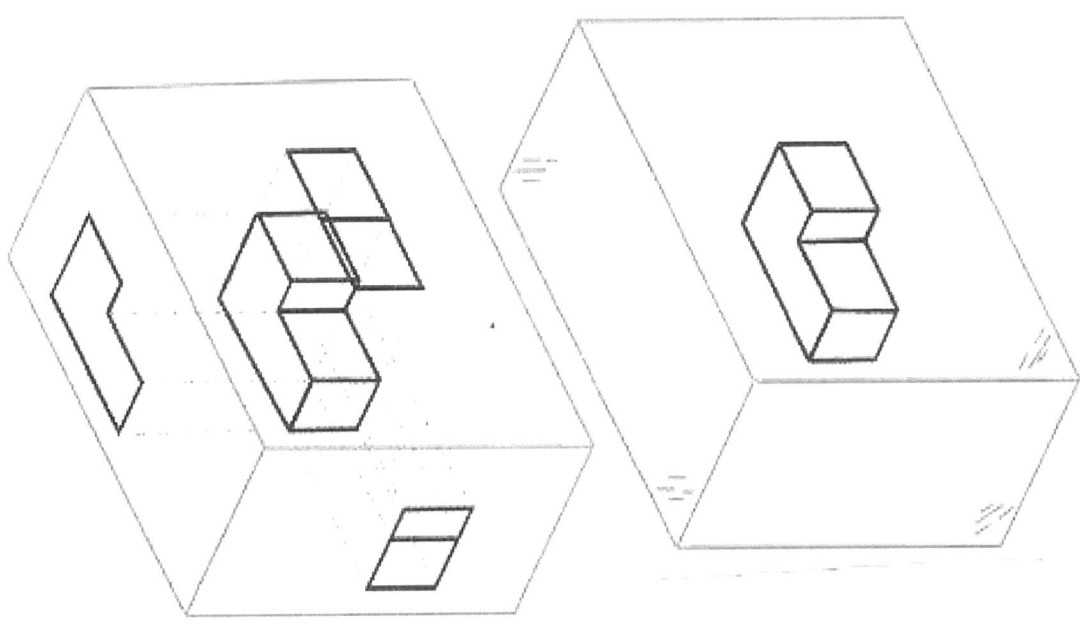

Every accelerated mass sees the universe ahead as a stack of infinite inertial planes. But because the reactionary fields of the planes are perpendicular thereto, the counter flux field lines of all planes are antiparallel to the acceleration – ergo accelerated bodies experience the inertia of the universe as though it were a single omnipresent infinite plane σ_U. Static empty space, however, cannot act as a transmission medium -- how is it, that accelerated mass feels the inertial impedance of distance matter? There being no physical contact, there can be no force without a dynamic deviser [a perplexity for Both Newton and Einstein inasmuch as both Newtonian and General Relativity Theory, require the injection of a mysterious acceleration constant G to actuate the equations]. As to be further elaborated, G is neither a constant nor a mystery, but rather an encryption of exponential spatial growth in relation to global mass. As the enabling source of isotropic spatial acceleration, cosmological expansion ubiquitously communicates the perpetual presence of inertial mass.

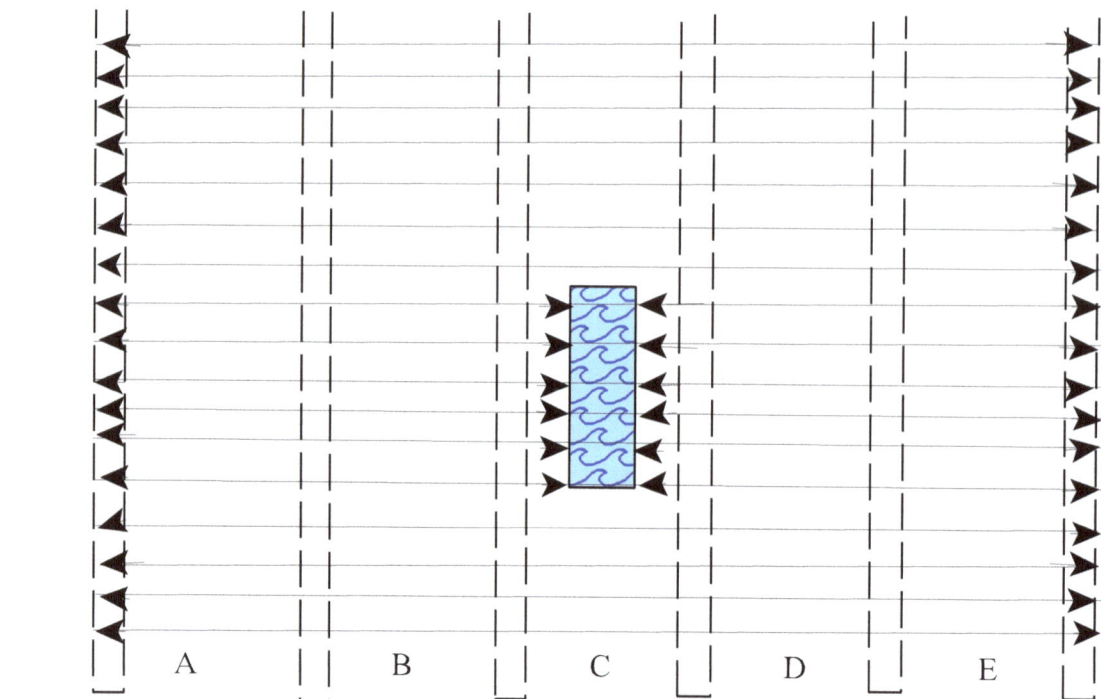

Fig 4. Expanding space shown as diverging arrows, inertial slabs A - E shown separated by dotted lines, the earth squashed to a flat blue disk viewed edge-on in C slab. As shown by the arrows convergent upon the earth's disk, the free expansion of inertial space is impeded by the earths inertial mass. Expanding space creates equal and opposite pressures within the cookie-cutter area defined by the projection of the earth's surface upon the area density σ_U of the combined cosmic slabs:

$$\frac{a_n}{g} = \frac{\sigma_U}{\sigma_B} \tag{1.9}$$

Thus, if gravity were unidirectional and the earth could be transformed to a flat surface of density σ_E = M_E/A_E, then (1.9) would be a correct expression for the earth's 'g' field. The reality of the situation is much different. Nonetheless (1.9) gives the correct result.

Fig 5, illustrates the actual operative cosmic for 3-D expanding space. Inertial slabs have been replaced by inertial shells, but the concept of inertial space as a scalar dynamic σ_U modulus is retained. Each shell has the same area density, consequently the collective of space from earth's perspective increases linearly with distance as measured from the earth's center. The intensity of the earth's spatial pressure field 'g' thus falls off inverse squared. At the Hubble limit **R**, the total density encountered by the expansion of inertial space (that accumulated by all inertial shells between earths surface radius r_e to the Hubble radius **R**, equals σ_U.

The projection of a pie shaped section of earth extended to the Hubble limit, illustrates how the intensity of the expansion pressure dilutes with distance. All cosmic mass can thus be considered distributed over the Hubble surface at distance **R**. As is the case for unidirectional accelerations, isotropic expansion creates equal and opposite forces between the universe and its constituent parts.

Taking the earth as an approximately uniform spherical mass density σ_E, the surface pressure due to expanding inertial space is:

$$P_E = (a_n)(\sigma_E) \tag{1.10}$$

and the pressure on the universe is:

$$P_U = (a_2)(\sigma_U) \tag{1.11}$$

For net momentum flow = zero, the two pressures will be equal, therefore:

$$a_2 = a_n \frac{\sigma_E}{\sigma_U} \tag{1.12}$$

FIG 6

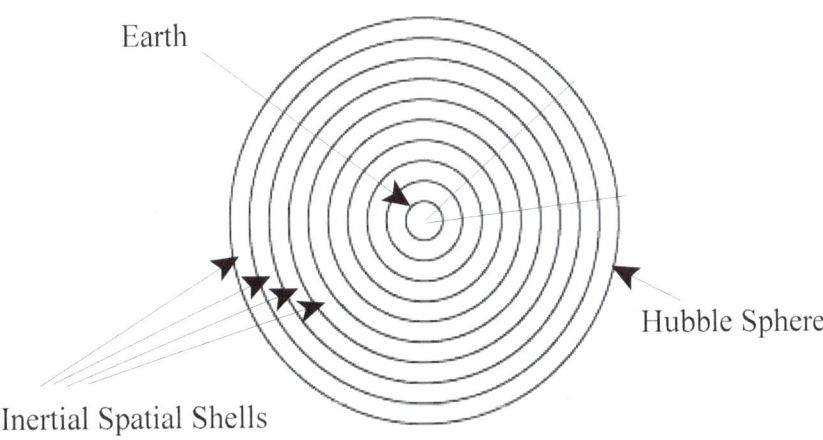

Earth

Hubble Sphere

Inertial Spatial Shells

Fig 6. All shells have the same 3-D density ρ_U and consequently equal area densities. Expansion creates two pressures [convergent momentum flux (called "gravity")], and [divergent momentum flux (Hubble recessional flow)]. While even the most dense materials are highly porous to spatial flux, matter nonetheless impedes spatial flow, creating thereby, the '**g**' fields of the masses.

To find the pressure created by expanding inertial space, we approximate earth as a functionally equivalent uniform spherical shell σ_E fashioned from a volume to surface transformation of its matter content. Likewise the operative essence of the Hubble as an impedance is defined in terms of its surface area density $\sigma_U = M_U/4\pi R^2$. The inertial relationship between the earth and the cosmos thus simplifies to **Fig 7**. All spherical masses as well as the universe, can be considered shells for purpose of calculating '**g**' fields. However, as elaborated below in the discussion of (1.18), account must be taken for the mass deficit when transforming Hubble energy content to a shell construct.

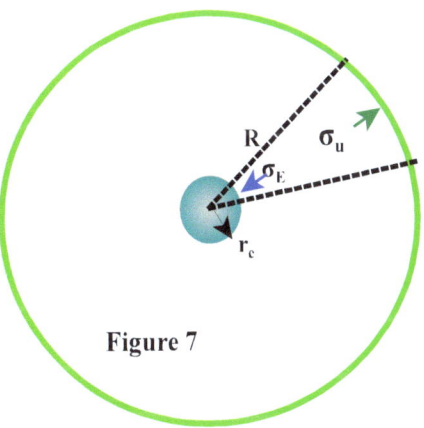

Figure 7

The explanation of inertial reactance as a scalar area density revisited as spatial inflow. In this scenario, cosmological expansion creates positive *a la* the acceleration factor (by whatever name **G**, **ΛR**, $\mathbf{a_n}$ or $\mathbf{c^2/R}$), which converges inwardly upon the shell area density of a local mass (e.g., $\boldsymbol{\sigma_E}$) and outwardly divergent upon the Hubble mass density ($\boldsymbol{\sigma_U}$) to create an inwardly directed acceleration flux *a la* Newton's 3rd Law.

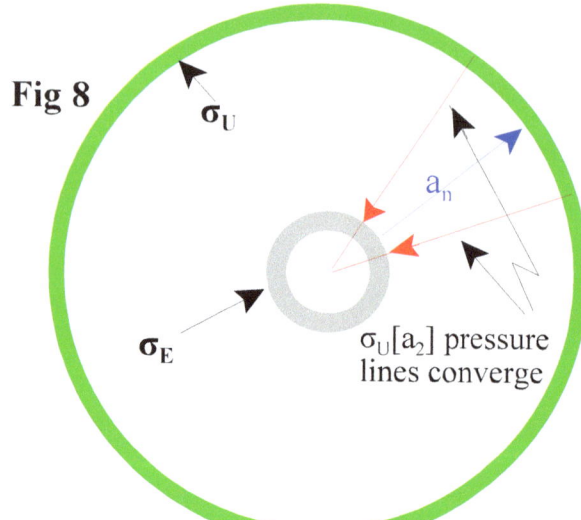

Fig 8

σ_U

a_n

σ_E

$\sigma_U[a_2]$ pressure lines converge

From Newton's Law $\mathbf{g = M_B G/r^2}$

From (1.16), $\mathbf{a_2 = g = \sigma_E a_N/\sigma_U}$

Whence $\mathbf{G = c^2/4\pi R\sigma_U}$

$\mathbf{R = 1.08 \times 10^{26}}$ meters, $\boldsymbol{\sigma_U} = 1$ kg/m^2

$\mathbf{G = 6.7 \times 10^{-11}}$ m^3/sec^2/kg·

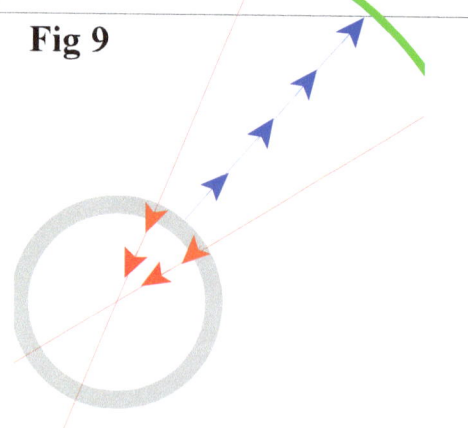

Fig 9

The Force on the earth $\mathbf{F_E}$ created by expanding inertial space pushing the universe outwardly at an accelerating rate $\mathbf{a_n}$ against the inertial resistance of the universe $\boldsymbol{\sigma_U}$ is equal and opposite to the force on the universe $\mathbf{F_U}$ created by the earths resistance $\boldsymbol{\sigma_E}$. Symbolically:

$$\mathbf{F_U = F_E = (M_E)(a_n)} \qquad (1.13)$$

Normalizing both forces in terms of one square meter of area:

$$\frac{\mathbf{F_U}}{\mathbf{m^2}} = \frac{\mathbf{F_E}}{\mathbf{m^2}} = \frac{\mathbf{M_E a_n}}{\mathbf{m^2}} \qquad (1.14)$$

Expressing the cosmic force $\mathbf{F_U}$ as 2nd law acceleration pressure:

$$\frac{\mathbf{M_U[a_2]}}{4\pi R^2} = \frac{\mathbf{M_E[a_n]}}{4\pi r^2} \qquad (1.15)$$

Hence:

$$\boldsymbol{\sigma_U}[\mathbf{a_2}] = \boldsymbol{\sigma_E}[\mathbf{a_n}] \qquad (1.16)$$

As previously(1.12), $\mathbf{a_2}$ is the '**g**' field.

The $\mathbf{a_n}$ and $\mathbf{a_2}$ acceleration forces do not terminate on the inertial shells $\boldsymbol{\sigma_U}$ and $\boldsymbol{\sigma_E}$ transformed from 3-D distribution. Rather, they act upon each shell inertial space proportionateley. From an inertial perspective, the difference between matter and inertial space is one of degree - both are mostly space. In this interlude, the notion of space as an inertial impedance is supported by fact -isotropic expansion in a curved Hubble environment leads to the same area density as the infinite plane slab model opposing unidirectional accelerations.

Gravitational field intensity diminishes inversely with the square of the distance 'r,' while for the same radius, area increases with distance squared. Consequently, force summed over the area $4\pi r^2$ is the same at every distance, so pressure abates inverse squared, and therefore the cosmic force upon $\mathbf{M_E}$ is equal and opposite to the force of $\mathbf{M_E}$ upon the cosmos. In words, a universe constructed from shells creates the same force as the inertial slab model of the cosmos. The effect of a shell at radius 'r' is the same as the affect of any other shell (all earth centered shells having the same thickness, irrespective of their radius from the earth, produce the same contribution to the "g" factor of the earth. The earth's area density $\sigma_E = M_E/4\pi(r_E^2)$, and the Hubble area density $\sigma_U = \mathbf{one\ kg/m^2}$. For the $\mathbf{q = -1}$ universe, $\mathbf{a_n = c^2/R}$ and "g" reduces to a ratio of surface densities σ_E/σ_u:

$$\frac{g}{A_n} = \frac{\sigma_e}{\sigma_u} = \frac{\dfrac{M_e}{4\pi(r_e)^2}}{\dfrac{M_u}{4\pi(R_U)^2}} = \frac{\dfrac{M_e}{4\pi(r_e)^2}}{\dfrac{kg}{meters^2}} \tag{1.17}$$

For earth: $\mathbf{M_E = 5.98 \times 10^{24}\ kg}$
$\mathbf{r_E = 6.37 \times 10^6\ meters}$

$\mathbf{g_E = [M_e/4\pi(r_E)^2](c^2/R)(meters^2/kg)}$

$\mathbf{= [1.173 \times 10^{10}\ kg/m^2](9 \times 10^{16}\ m^2/sec^2)/(1.08 \times 10^{26}\ m)}$

$\mathbf{= 9.8\ m/sec^2}$

For the earth, Gauss's law closely approximates the fidelity of surface density σ_E as the alter representative of volumetric density ρ_U. However, for masses and sizes that result in significant non linear effects (gravity acting upon gravity to create more gravity), corrections are required. For the Hubble sphere, the difference between the gravitational energy of a 3-sphere and a 2-sphere will be zero when $\mathbf{R_2 = (5/6)R_3}$.

$$U_3 - U_2 = \frac{3M_U^2 G}{5R_3} - \frac{M_U^2 G}{2R_2} = 0 \tag{1.18}$$

Fig 9 properly reflects the action of the expanding universe within the volume contained by the Hubble area $\mathbf{4\pi R^2}$ where $\mathbf{R = (5/6)R_H}$. The Hubble radius $\mathbf{R_H}$ is approximately 1.3×10^{26} meters for $\mathbf{H_o = 70}$. The radius of the 2 sphere emulative that will produce the same gravitational attraction is therefore 1.08×10^{26} meters. When the measure of the earth's 'g' field is taken at its surface, it is predicated upon both the spatially impeding mass of the universe σ_U and the spatially impeding action of the earths inner structure as represented by the surface density σ_E of the earth. Like most astronomical bodies, the earth's mass and size are such that no adjustment is required to account for the change in binding energy when transforming from a 3-sphere to 2-sphere.

As far as acceleration and gravity are concerned, the impedance of the universe is isotropic - acceleration is opposed equally in all directions because the universe is homogeneous on the large scale. Impedance emerges as the consequence of changing momentum, in form as infinite slabs of infinite area. Yet impedance is non existent without some relative acceleration between the M_B and the universe. Functionally, the cosmic modulus acts as an instantly adjacent area density in any direction in which M_B is accelerated. Although cosmic mass is distributed throughout the universe homogeneously on the large scale, resistance to acceleration is explained metaphorically within the fiction of a large number of low density slabs acting in concert. In terms of Hubble perimeters, the cosmic scale R is on the order 10^{26} meters. Reworking the problem by taking a sample volume in the form of a 10^{26} meter cube containing a total bare mass energy M_b, the area density function in each of the three dimensions is:

$$\rho_U = \frac{M}{L^3}, \ therefore$$

$$\sigma_U = \frac{M}{3L^2}, \ hence \tag{1.19}$$

$$\rho_U L^3 = 3\sigma_U L^2$$

$$From \ which \ \sigma_U = \rho_U \frac{L}{3}$$

Taking Hubble density as **3 x 10^{-26} kg/meter³**, then for **L = 10^{26}** meters σ_U = **one kg/meter²** The ratio of Hubble volume to Hubble area is:

$$\int_V \rho_U \, dV = \int_S \sigma_U \, dS$$

$$\rho_U \left[\frac{4\pi R^3}{3} \right] = \sigma_U \left[\frac{4\pi R^2}{1} \right] \tag{1.20}$$

$$whence \ \sigma_U = \rho_U R/3$$

Which has the same form as (1.19). The scalar density calculated from the cube will be slightly different from that calculated for the cube owing to the difference in the volume to surface area ratios between a cube and sphere. To check our calibration for $\sigma_U = 1$, we used the earth as a known gravitational field (1.17). From (1.10) and (1.11) our estimate of the inertial density of the void $3\sigma_U/R$ should comport the energy density of the vacuum. More specifically, for the zero energy universe,

$$-P = (\sigma_U)(a_N) \tag{1.21}$$

$$\rho_U = \frac{-3P}{c^2} = \frac{-3(\sigma_U)(a_N)}{c^2} = \frac{-3\sigma_U}{R} \tag{1.23}$$

which for $\sigma_U = 1$ is consistent with our initial estimate of cosmic density (**3 X 10^{-26} kg/m³**).

SIDELINES ON GRAVITY AND INERTIA

The following pages repeat much of what was preciously
presented and is therefore mostly redundant. It represents
an earlier development focusing upon gravity as a reactionary
force in massless space.

Sidelines On Gravity ----Inertia Revisited and Rehashed

Theories of force based upon momentum exchanging particles (virtual photons, gluons and gravitons) have no predictive power.[25] That the constabulary embraces regiments without reasons, works in the end, to the suppression of what Einstein called "the holy grail of curiosity." Ideas, once ingrained, and endorsed by pier approval, are not easily undone by logical assault upon validity. Still, scientific reforms occur, often brought about by an *"upstart crow"* with the temerity to interrogate treasured doctrines.

Experts positioned to bring about meaningful change, however, are hampered by the prospect of losing credibility with colleagues when they stray too far from standard theory. Commonly held beliefs are seldom scrutinized with an eye to a complete overhaul. While modern methods reveal fallacies in long cherished beliefs, flawed theories and/or unproven parts of theories, persist as proven postulates.[26] In these pages, the principles considered sacred are those that have survived scrutiny in the empirical court of experimental confirmation. A theory will be considered correct only to the extent it has been tested in the manner required to satisfy the breadth of the hypothesis.[27] task of expressing natures forces in terms of cosmological parameters, is greatly simplified for a zero energy balanced universe. When properly interpreted, the agency of self creating expansion, reveals both the means and manner by which gravity evolves. In the end, that which provokes the divergence of masses on the cosmological scale will be understood as one-in-the-same, as that which causes masses to converge on a lesser scale.

Gravity is predicated upon symmetry of action, specifically, Einstein's *"Principle of Relative Acceleration."*[28] Newton's *'Law of inertia'* and his *"Law of gravity"* are unified within Einstein's *Principle Of Equivalence,* herein taken to a new level of simplicity. Inertial reaction created by accelerated mass becomes indistinguishable from gravity fashioned from spatial acceleration.

For a body **B** having a fixed inertial mass M_B, the reactionary force for an acceleration (**a**) is:

$$\mathbf{F} = \mathbf{M_B(a)} \tag{1}$$

---

[25]The idea that forces are conveyed by momentum transferring particles can, in part, be attributed to a theory put forth by Hideki Yukawa in 1933. Reasoning that the Coulombric repulsion between electrons, if multiplied by a factor reflecting a temporal conflation delay of approximately 200, would correspond to a particle mass in the range of 200 electrons. The *"mu meson"* ($207m_e$) was originally thought to be Yukawa"s particle, but as later turned out, it did not fit the quantum theory that had evolved around the idea of spin 1 quantums traveling between nucleons. Later, the somewhat heavier *'pion'* was discovered and adapted to fulfil the need.

[26]Specific examples include 1) gravity as curved static space, 2) the failure to heed the well established fact that mass is not a conserved quantity., 3) Likewise, the constancy of **G** is mis-established from conclusions drawn from orbital studies, the stability of which evidences only the constancy of the **MG** product.

[27]*Remodeled Relativity Theory* - A revision of Einstein's two Relativity Theories incorporating only experimentally proven principles. Abhijit Biswas, Krishnan RS Mani

[28]No difference exists between acceleration of mass wrt space and acceleration of space wrt mass.

Normalized in terms of a common area of one square meter:

$$\frac{F}{kg} \times \left[\frac{kg}{m^2} \right] = \frac{M_B}{m^2}(a) \qquad (2)$$

Then

$$\frac{F}{kg} = \frac{\sigma_B}{\sigma_u}(a) \qquad (3)$$

Where $[kgm/m^2]$ is denoted as σ_U and M_B/m^2 as σ_B. Equation (3) states Newton's 2nd law in terms of reactionary *"force-per-kg"* created by an acceleration 'a' applied to σ_B. The left side of (3) is reactionary force per kg (dimensionally an *acceleration* having **SI** units of *meters/second²*). Hereinafter, reactionary accelerations are labeled '**g**,' (the consequence of the implacable demand that momentum be conserved on the cosmic scale). Newton's 2nd Law expressed in terms of pressure(s) is thus:

$$g(\sigma_U) = a(\sigma_B) \qquad (4)$$

which will be understood as a restatement of Pascal's law, specifically, $(\sigma_B)a$ must be balanced by a counter pressure $(g)\sigma_U$. The restatement of Newton's 2nd law [(3) and (4)] in terms of pressures leads to leads to a new appreciation of gravity in the context of inertia -- to balance the source pressure $a(\sigma_B)$, the reactionary acceleration '**g**' must be multiplied by a fixed factor σ_U which, to be compatible with the dimensional formalism, must equal *one kg/m²*. Pressure is momentum flow, the force between the universe and an accelerated body is defined by momentum flow.[29] What then is the physical meaning of σ_U?

It is the surface density of an operative inertial plane of infinite extent. While the Hubble sphere does not act as an inertial entity, in and of itself, it does provide a convenient sample for estimating average volumetric density of the universe as a whole. To find the density of the infinite plane, Hubble measurements of density is transformed to a shell *a la* the divergence theorem and thence to a plane, taking into account deficits that occur when masses are transmuted between different geometries.[30]

Reasonable estimates of Hubble size and bare mass (e.g., from Appendix I, R_H = 1.3 x 10^{26} meters and M_U = 1.5 x 10^{53} kg), comport with the requirements needed to create a one **kg/m²** surface density. To satisfy Newton's 2nd law, however, the operative value σ_U of the infinite inertial plane, must be exactly one **kg/m²**.

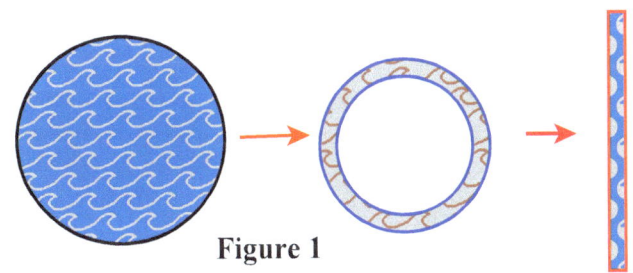

Figure 1

---

[29]Pressure is momentum flow (**mv/sec**)

[30]For a derivation of the 2-sphere density based upon the estimates of Hubble size and mass, see Appendix I.

In this expose' *"Einstein's Equivalence Principle,"* takes center stage, not as the cause of curvature, but rather as the inertial reactance of matter opposing volumetric spatial acceleration.[31] Gravity has remained a mystery because '**g**' fields are incorrectly perceived as created by matter rather than the reaction of matter.[32] Inertial reactions developed in response to the isotropic field of expanding space are '**g.**' fields – issued as counter accelerations pursuant to the mandate that momentum be conserved on the global scale. Formulated as pressure per (4), gravity conceptualizes as momentum flow. Every mass demands a mass proportionate '**g**' field momentum influx to offset the reactive force created by the action of expanding space upon non-expanding matter[33]

Delineating σ_U in its operative role as the inertial essence of the universe might in one sense be objectified as an ether.[34] It is foundational to the unification of inertia and gravity. To create inertial reactions, acceleration is required. Mass alone creates no force, nor does it curve space without the aid of **G**. Indeed, Einstein's hypothesized curvature of space as the alternative to gravity as force, nonetheless, requires **G**).[35] But the existence of an isotropic acceleration field abrogates need for further postulation. Gravity follows straightaway, the auspices of Newton's 2nd Law. The infinite plane, in and of itself, creates no force. In representing the reactionary action of the universe, it does so as a facsimile. To emulate the force, acceleration is added. The dynamic essence of the infinite plane does not emerge until spatial acceleration is added, consequently no attractive nor reactive force arises between the mass of the universe and its individual masses[36] Both Newton and Einstein relied upon an empirically determined **G** factor to proved the acceleration, but neither man could explain its cause. Herein, expansion is commandeer to provide the source of isotropic spatial acceleration. That expatiation of the void comports with what is required to derive **G** from '**g**' fields, leads to the following: From (4):

$$\mathbf{g} = \frac{\sigma_B}{\sigma_U}\mathbf{a}_n$$

---

[31]The Equivalence Principle does away with the idea of a separate gravitational force.

[32]Saith Richard Feynman: *"Gravitation is a new field of its own ....or a consequence of something already known but incorrectly perceived "* (Feynman - Lectures on Gravity).

[33]At the subatomic level, particles are held together by electrical and quantum forces. Nearby masses resist separation because they are gravitational bound (ironically the mechanism that causes masses to be mutually attracted at close distances is one-in-the-same as that which causes galaxies to separate on the cosmological scale.

[34]The mathematical migration of the Hubble contents to an operative inertial plane consistent with the measured relationship between force, mass and acceleration, was unknowingly brought about by the early experimenters following Newton's discovery of the 2nd Law. By defining the '**ntn**' as the unit of force corresponding to (*one meter per sec² per kg*), they, in effect, assigned the dimensionality and magnitude of the infinite plane σ_U as one **kg/m²**.

[35]The excess radius $\mathbf{R}_{ex} = \mathbf{MG/3c^2}$

[36]As is the case with both Newton's law and GR, the equations are impotent until the acceleration factor G is introduced.

In (5), 'a_n' is the cosmological acceleration c^2/R created by expansion. To represent the state of the Hubble universe in terms of its acceleration, velocity and scale, the misnamed deceleration parameter 'q' will have a value of (-1) for an exponentially expanding universe, hence, if R is the effective Hubble scale for a 2-sphere having the bare mass content of the Hubble sphere, then from (P-7):

$$q = -1 \frac{\ddot{R}R}{\left(\dot{R}\right)^2} \tag{6}$$

And therefore:

$$\ddot{R} = \frac{\dot{R}^2}{R} = \frac{c^2}{R} \tag{7}$$

Hence:

$$g = \frac{M_B}{A_B}\frac{1}{\sigma_U}\left[\frac{c^2}{R}\right] = \frac{c}{4\pi r^2}\left[\frac{M_B}{R\sigma_U}\right] = \frac{M_B}{r^2} \times \left[\frac{c}{4\pi R\sigma_U}\right] \tag{8}$$

From Newton's law of gravity:

$$g = \frac{M_B G}{r^2} \tag{9}$$

Then from (8) and (9):

$$G = \frac{c^2}{4\pi R\sigma_U} \tag{10}$$

previously arrived at (P-9). If (8) and (9) correctly define 'g,' then (10) necessarily encodes Einstein's ΛR Factor = c^2/R. That (10) produces a plausible expression for G, is based upon the operative value of R required to account for the mass deficit (loss of binding energy incurred in transforming from 3-sphere to 2-sphere per Appendix III). Accordingly, if the Hubble scale R_H is 1.3×10^{26} meters (estimated "*now*" value of the Hubble parameter H_o =70), then the operative value of R is 1.08×10^{26} meters, whence:

$$G = [6.6 \times 10^{-11} \text{ m}^3/\text{sec}^2][\text{kg}]^{-1} \tag{11}$$

In the forgoing, Newton's 2nd law is embellished to express inertial reactions as pressures, hence momentum flow. In the formulation, force and mass were normalized *a la* an artificially created common surface area (**one meter²**). Thus in (2), the density σ_B is that obtained by dividing the mass of a body '**B**' undergoing acceleration by an area representing a cross section defined in terms of the geometry of the body in relation to the direction of the acceleration (an easy problem for a uniform density flat sheet accelerated normal to its surface, but difficult to assess for odd shaped nonuniform densities).

When applying Newton's 2nd law in the context of total reactionary force, there is generally no need to know the distribution of the inertial reactionary field within the interior since it has the same distribution as the weight. The determination of 'g' field force density on the surface of astronomical bodies is frequently of interest - for present purposes the interest will be in checking the result predicted by (5) against measurements arrived at by other means. As previously elaborated, the effective area of σ_B for 2nd law purposes is independent of the orientation of the body or the distribution of its mass because the influence of the contrived infinite plane is always apposite to the direction of the acceleration and independent of the distance between the body and the plane. Unit areas of *force/m²* (wherever found) on the left side of (2) thus match unit areas of *mass/m²* (wherever found) on the right side of (2). That σ_U density must be **one kg/m²** will be understood by reverting the pressure equation back to a force equation (2) for an unknown mass [?].

$$\frac{F}{kg}\left[\frac{?}{m^2}\right] = \frac{M_B}{m^2}(a) \tag{12}$$

Since {*Force per kg*} is dimensionally an acceleration, then [?/m²] must have units of [**kg/m²**] in order that the left and right sides of (12) be in dimensional balance. But since **m²** in the denominator of [**?/m²**], cancels **m²** in the denominator of M_B/m^2, (12) reverts to Newton's 2nd Law if the unknown mass factor [?] equals **one kg** and σ_U = **one kg/m²**. Since both σ_B and σ_U are expressed in units of **kg/m²**, the ratio of the densities (σ_B/σ_U) determines the ratio of the accelerations **g/a**.

The dynamic response of the universe to an upward acceleration of the **L** shaped structure is shown in **Fig 2**. Counter reaction is brought into play by the self adjusting orientation of the infinite plane σ_U (depicted edge on in Fig 2). Counter pressure created by σ_U upon internal elements of L constitutes the reactionary force **F**. Because The magnitude of the counter acceleration acting upon any element of mass is proportional to the volumetric mass density defined by the surface area per unit volumetric unit of mass

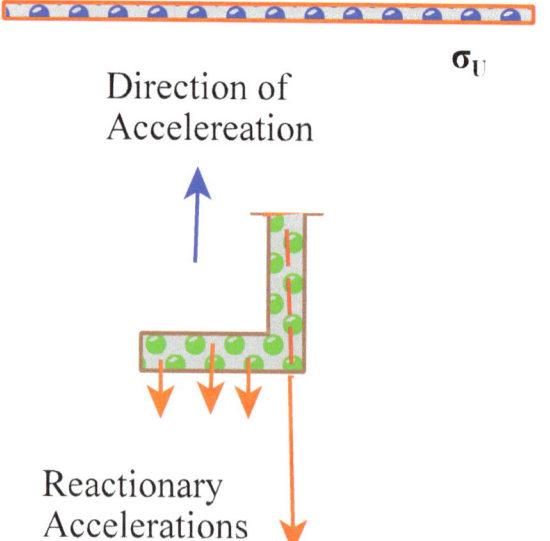

Direction of Accelereation

σ_U

Reactionary Accelerations

While the experiment says nothing new about gravity-inertial equivalence, the infinite plane as *modus operandi* proffers the proposition that an extended agency makes its presence known in the form of inertial reaction when masses are accelerated. As a uniform scalar density distribution of cosmic matter, infinite planes create parallel inertial reactive forces consistent with the emplacement of matter within structures. The implication of distended dynamic reaction is fortified by the observed action of reactive forces upon individual masses irrespective of the binding forces that define the structure.

Infinite Plane —> Dynamic Membrane ---> instantaneous reactance

Fig 3 illustrates a possible juxtaposition of infinite planes that arise when 'L" is accelerated upwardly. Total reactionary force is the sum of the elemental reactions. Because reactionary planes can be considered anywhere and everywhere with respect to accelerating mass, reactionary forces are instantaneous. The universe gets it right, a demand imposed by the intractable laws of conservation of momentum and energy.

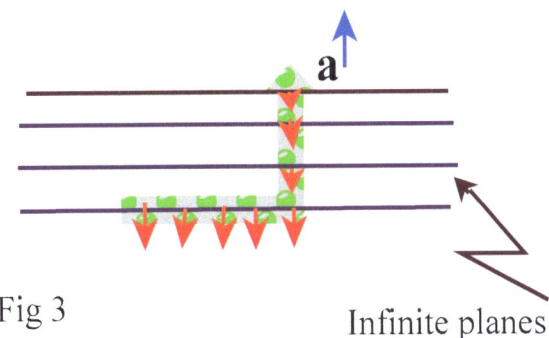

Fig 3

Infinite planes

Infinite planes functionally exist where ever located. A single ∞ plane of density σ_U can be represented by a number of lesser planes each having appropriately reduced density such that the total density of all planes equals σ_U. In Fig 3, the planes are depicted as collinear with the mass centers of the inertial elements comprising 'L'

Mass plays multiple roles - e.g., *momentum, kinetic energy, rest energy, force, gravity* .... all defined in terms of a single coefficient **M**, which is neither constant nor conserved. How is it that the inertial property of matter arises? Herein, the development of mechanism will embrace Machianism. The idea that distant matter determines local inertia, first appears in a 1721 work by Bishop Berkeley. In rejecting Newton's idea of an absolute space, Berkeley proposed that local inertia was due to a sky of stars. As previously related in the Prelude, the 19th Century Physicist , Ernst Mack, proposed a similar theory based upon the influence of matter scattered throughout the universe. "Mach's Principle," has survived as an intriguing proposition despite criticism.[37] Arguendo, the concept of the infinite plane disposes of Einstein's remonstrance. It is nothing short of remarkable that the infinite plane acts at once to dispels protestations founded upon SR -- by virtue of the fact σ_U has no effect upon the value of the masses that create the '**g**' fields. Thus, while the totality of cosmic mass-energy emerges instantly to oppose acceleration, it the similitude of an infinite plane, σ_U has value unity, and therefore has no influence upon the magnitude of the reactionary force (which depends solely upon the mass of the entity subjected to acceleration).

Mach's Principle, reinstated in form as a dynamic scalar operative, abrogates cosmological inconsistencies. The starting point for correctly remodeling the universe begins with the fallacious idea that inertial mass is constant. That **M** is neither conserved nor constant is a known fact - yet the notion of inertial as an early-on acquisition of particles, stubbornly persists.

---

[37] "....*the inertial resistance opposed to relative acceleration of distance masses presupposes* [instantaneous] *action at a distance; and as the modern physicist does not believe he can accept this action at a distance, he comes back once more to the ether ...."* Excerpt from Einstein's 1920 address at Leyden university. Einstein rejected Mach's Principle because it appeared to require faster than light communication.

The idea of gradually acquired inertia is admittedly difficult to accept.[38] Unlike the defunct "*Steady State Theory*" and the hypothesized "*inflation theories,*" (where density is maintained constant during expansion by creation of new particles) the cosmo-dynamic theory of gradual inertia enhancement requires no new particles. That the inertia **M** of existing particles $\propto$ **R**, is justified by the rate at which negative energy 'g' fields dilate during expansion. All mass exhibiting particles are accompanied by a negative energy 'g' field coextensive with the volume of the universe. Negative energy 'g' fields exists as spatial stress (tension). A 'g' field having stress energy **E** has mass **E/c²**. The **E/c²** mass contained in the 'g' field of all the particles contributes approximately 50% to the inertial reactance of individual particles. Volumetric spatial expansion increases total negative energy. Doubling cosmic sized doubles cosmic negative energy, doubling the negative energy in the gravitational field of a particle doubles the positive **Mc²** energy of the particle. Consequently cosmic inertial energy increases $\propto$ **R²** per (P-15) That is why the inertial characteristic of existing particles increases during expansion.[39]

As introduced in the prelude, the first theoretical formulation of an expanding universe was discovered in 1917. Having neither pressure nor density, de Sitter's empty universe appeared to have little practical application.[40] That would change as a result of later discoveries. Friedmann's 1922 development of a density dependent expansion formalism was the beginning of the end for the static universe. Shortly thereafter, Edwin Hubble published data connecting red shift and luminosity, which led to the velocity-distance law (now known as the expanding universe). Comes then English cosmologist, William McCrea, with a theory of the void founded upon tension. Interest in De Sitter's exponentially expanding empty universe should have been aroused, as it could have been applied to McCrea's real universe to explain the expansion profile when positive matter energy equals negative pressure energy. Indeed, the solution for the zero energy state of McCrea's universe corresponds precisely with de Sitter's "*Exponentially Expanding Empty Universe.*" According to McCrea's opus, expansion of a negative pressure volume creates positive energy. Expansion is self sustaining as well as self initiating. And as shown in the prelude, no dark energy is required.

An expanding zero energy universe remains balanced because a dilating negative pressure space simultaneously creates positive energy while negatively energizing the newly created space. Empirical support for the exponentially expanding proposition, however, would await the 1998 **1a** supernova studies.

---

[38]Whether in the form of an unexplainable 'big bang' beginning, or a short interlude of *ad hoc* inflation, for some obscure reason, matter gets priority over space and time in creation theories. Ideas once ingrained, are not easily disposed. Perhaps it is the mental attachment in the mind of humans that yearns for the comfort of permanence attributed to the material world upon which all living things owe their existence.

[39]The issue of whether inertia varies with velocity, depends upon how one views relativistic mass. In Einstein's formulation, it is the manifest property of inertial matter that increases with relative velocity. For others, it is the effect of time dilation. The passage of time, and consequently the temporal rate of clocks, however, depends upon the relative energy state in the same way that time depends upon gravitational potential. And since gravity is an inertial effect, the two explanations ultimately reduce to a common factor.

[40]De sitter proof, Appendix III.

The 1934 derivation of the gravity equations by Milne and McCrea, reduced GR to a simple energy relationship.[41]

$$\ddot{R} = -\frac{4\pi G}{3}\left[\rho_U + \frac{3P_s}{c^2}\right]R + \frac{\Lambda R}{3} \tag{13}$$

Equation (13) is identical to (P- 20), which, as previously, reduces to de Sitter's solution when:

$$\rho_U c^2 = -3P \tag{14}$$

For the zero energy state, positive inertial energy $(\rho_U c^2)/3$ equals negative pressure (-P), ergo equation (14) reduces to:

$$\ddot{R} = \frac{\Lambda R}{3} \tag{15}$$

To avoid gravitational collapse, Λ must equal $3H^2$ and therefore:

$$\ddot{R} = \frac{3H^2 R}{3} = \frac{c^2}{R} \tag{16}$$

which will be recognized from (6) as the definition of the $q = -1$ deceleration parameter. To reiterate, the universe expands exponentially when positive mass energy = negative gravitational energy, consequently it needs no help in the form of "*dark energy*." The zero energy state is sustained during accelerated expansion, the funding thereof being at once due to the increase in positive energy created by expansion of a negative pressure volume. An equal increase in negative gravitational energy follows from the inertial enhancement of **M** being then spread throughout the expanded volume. Positive and negative energy jointly contribute to the rate of growth.

Inertial reaction and gravity are alter identities. The intensity of '**g**' fields are maintained throughout the universe by spatial expansion. Any change in the intensity of a gravitational field brought about by acceleration of the mass **M** of which it is a part, will be communicated to the infinite plane for opposition. In this way, existing '**g**' fields are distorted.[42] Inertial reactance is no less inculpated as negative energy contained in the '**g**' field of a mass. Expanding space communicates the acceleration of a mass to the σ_U plane which creates a change in the existing '**g**' field as response. An amendment to the gravity field heralds a new acceleration. As gravitational forces can be derived in terms of 2nd Law Newtonian reactions, so also inertial reactions can be understood as '**g**' field distortions.

---

[41]Appendix IV

[42]For example, if the earth were accelerated at 9.8 meters/sec along the direction of its rotational access, polar bears would weigh twice as much and penguins would be weightless.

From (4), one obtains for the earth:

$$g = \frac{\sigma_E}{\sigma_U} a_n = \frac{\dfrac{M_e}{A_E}}{\sigma_U}[a_n] = \frac{\dfrac{5.98 \times 10^{24}\ kgm}{4\pi(6.37 \times 10^6\ meters)^2}}{\dfrac{kg}{meter^2}}\left[\frac{c^2}{R}\right] = (1.173 \times 10^{10})\left[\frac{c^2}{R}\right] \qquad (17)$$

Using the value of $R = 1.08 \times 10^{26}$ meters, defined by the infinite plane,

$$\textbf{g = 9.8 meters/sec}^2 \qquad (18)$$

From a geometric perspective, exponential expansion corresponds to flat space The mathematical description of the universe simplifies from (13) to (15). The mathematical properties of an operative infinite plane become reality in a physically infinite universe. There is no need to define space with greater certitude or specificity than as warranted by the experiments. Indeed, space eludes detection when at rest, but gravity fields akin to inertial reactions, follow from the acceleration field created by cosmological expansion. As in the prelude, two things are known about space [43]

In its original form, Newton's Law of Gravity, was deduced as an inverse squared acceleration field required to explain Kepler's laws of planetary motion. The value of G was later measured separately using a torsion balance with known masses. Establishing G as a separate and distinct cosmological property having *mks* dimensionality *"cubic meters per second squared per kg"* narrows the mystery of gravity to what-in-the-universe is defined in the context of an accelerating volume. The answer in part, incorporates another mystery, that of why the ratio of two accelerations formed from different cosmological parameters, should equal one within the limits of experimental error:[44]

$$\frac{M_U G}{Rc^2} \approx 1 \qquad (19)$$

Where M_U is the mass of the Hubble sphere, and R is the effective radius. Herein, (19) is viewed not as a mystery, but rather as a definition of G. In words, G is the ratio of the volumetric expansio rate of the Hubble sphere divided by its inertial mass:

$$G = \frac{Rc^2}{M_u} \qquad (20)$$

---

[43] The pressure P of the vacuum (P-24) is the product $(\sigma_U)(a_n) = -8.8 \times 10^{-10}$ ntn/m^2

[44] The relationship was extensively studied by Carl Brans and Robert Dicke in their intensive search for a scalar tensor theory of gravity. Robert Dicke proposed the relationship should be recognized as the equivalence connective between inertial and gravitational mass by way of Mach's Principle. The problem boiled down to equation (10), namely that of deriving G in terms of R and c.

And since $\sigma_U = M_u/4\pi R^2$ then

$$G = \frac{Rc^2}{4\pi\sigma_U R^2} = \frac{c^2}{4\pi R\sigma_U} \qquad (21)$$

Which corresponds to (10).

Recognition of volumetric acceleration as the cosmological cause of the attraction between masses resolves the mystery of why the universe is nearly balanced between gravity and expansion.[45] Omega is one because $G = c^2 R/M_u$. The density of the cosmos is critical because the acceleration created by expansion is one-in-the-same as that which causes gravity.[46] In Einstein's final draft of General Relativity (1916-17), he introduced a counter acceleration field Λ (seemly required at the time to prevent gravitational collapse). For a static universe, Λ would need to equal $3H^2$ (which as it turns out is not constant in an expanding universe). When substituted into (13), the term $\Lambda R/3$ would according have the value:

$$3H^2R/3 = c^2/R \qquad (22)$$

The introduction of the cosmological constant Λ to balance gravity, in hindsight perfectly fits what is required of exponentially expanding space, to create gravity. Indeed, the isotropic acceleration field follows straightway as the ratio of Hubble volumetric acceleration divided by Hubble area, that is:

$$V = \frac{4\pi R^3}{1} =$$

$$\dot{V} = \frac{4\pi R^2}{1}\left[\frac{\dot{R}}{1}\right] \qquad (23)$$

$$\ddot{V} = 8\pi R(\dot{R})^2 + 4\pi R^2\ddot{R}$$

From (6), then:

$$\ddot{V} - 12\pi R(\dot{R})^2 \qquad (24)$$

Per Gauss's divergence theorem

$$\frac{\ddot{V}}{\text{Area}} = \frac{12\pi R(\dot{R})^2}{4\pi R^2} = \frac{3c^2R}{R^2} = \frac{3H^2R}{1} \qquad (25)$$

---

[45]Modern cosmology is understandably in awe of the fact that, by the best available evidence, the expansion rate is infinitesimally close to the critical density that corresponds to eternal decelerating expansion (gradually diminishes to zero at infinite length). Critical density ($\Omega = 1$) ceases to be much less of a cosmological wonderment with the realization that gravity can be explained and derived in terms of expansion.

[46]An expression for the critical density assumes Λ and 'k' zero in (13). When the substitutions are applied to Friedmann's formulation:

$$\rho_U = \frac{3H^2}{4\pi G} \qquad (26)$$

The value Einstein required to balance gravity in (13) corresponds to the volumetric rate of spatial expansion (25). Zero energy corresponds to exponentially expanding flat space.

$$\ddot{R} = \frac{3H^2R}{3}, \text{ whence}$$

$$\frac{\ddot{R}}{R} = H^2, \text{which has } solution \quad (27)$$

$$R = Ke^{Ht}$$

The advent of R in the denominator of (13) requires G decrease inversely with passage of time in a uniformly aging universe. While constancy of G is generally inferred from the stability of planetary and lunar orbits, the studies only attest to the invariance of the $M*G$ product.[47] For orbital stability, it is the $M*G$ product that must be constant. If the inertial property of matter increases during expansion proportionately with the scale factor R, the decrease in G will have no influence upon orbits. The temporal constancy of the product of any mass multiplied by the global G field defines a shared but hidden variance. To see how mass varies in a spatially expanding zero energy matter dominated universe, we consider the Hubble as a uniform density sphere having bare mass energy $(M_B)c^2$ and gravitational spatial energy $U_3 = 3M_B^2G/5R$. The total energy E_T for a radius R_H is:

$$E_T = M_Bc^2 - \frac{3(M_B)^2G}{5R_H} = 0 \quad (28)$$

Consequently:

$$M_B = \frac{5R_Hc^2}{3G} \quad (29)$$

For G constant during expansion, then the gravitational energy expressed in terms of the mass M_B must increase proportionally with R_H. Already we are in conflict with standard model precepts, but for the $M*G$ product of an existing central mass to remain constant during expansion as required for orbital stability, then G must diminish by half when R_H doubles.

Accordingly, the new value of M_B is:

$$(M_B)_2 = \frac{5}{3}\left[\frac{2R_H}{\dfrac{G}{2}}\right] = 4M_B \quad (30)$$

In words, during the time it takes for the Hubble scale to double, the gravitational mass-energy

---

[47]Where the central mass $M*$ is >> than a mass M_x following a nearly circular orbit, the expression for orbital parameters can be found by equating gravitational force to centripetal force, i.e., $M*M_xG/r^2 = M_x(v^2/r)$. Whence $M*G = v^2r$

increases as the square. Because the energy of any shell at a distance' $\mathbf{r}$' from the mass center is the same, doubling the radius of the Hubble sphere will result in an increase in total gravitational energy, that is:

$$\mathbf{U}_{2R} = \frac{3(4\mathbf{M}_B)^2 \frac{G}{2}}{5(2R)} = \frac{3}{5}\left[\frac{16\mathbf{M}_B^2}{2R}\right]\left(\frac{G}{2}\right) = \frac{3}{5}\left[\frac{4\mathbf{M}_B^2 G}{R}\right] \tag{31}$$

In general, whether modeled as a three sphere, 2 sphere, or flat plane, doubling of the scale factor $\mathbf{R}$ increases gravitational energy by $\mathbf{R}^2$. No new particles are assumed, and none are required -- the same number of atoms are distributed throughout the new volume which has increased $\propto \mathbf{R}^3$. And since the mass energy has increased $\propto \mathbf{R}^2$, the density ρ_U decreases $\propto 1/\mathbf{R}$.

Inasmuch as no assumptions were made as to the temporal constancy of σ_U, it now seen to be consistent therewith. That is, if $\mathbf{M}_u$ is the total effective mass of the universe, then from (2):

$$\mathbf{M}_U = 4\pi\mathbf{R}^2\sigma_U \tag{32}$$

The above relationships are consistent with what would be expected in the context of the increased negative energy that arises from the larger volume that contains the energy of the individual '$\mathbf{g}$' fields and their collective interaction. Total spatial energy equals Pressure times Volume, i.e.,

$$\mathbf{E}_s = \mathbf{PV} = -[\sigma_U]a_n\left(\frac{4\pi\mathbf{R}^3}{3}\right) = [\sigma_U]\left[\frac{c^2}{\mathbf{R}}\right]\left(\frac{4\pi\mathbf{R}}{3}\right) = \frac{4\pi\mathbf{R}^2 c^2}{3} \tag{33}$$

Spatial energy is thus proportional to $\mathbf{R}^2$, thence in a zero energy universe, positive $\mathbf{Mc}^2$ energy is also proportional to $\mathbf{R}^2$. Per (14), the zero energy mandate confirms the inverse dependence of density upon $\mathbf{R}$, specifically since $\rho_U = -3P/c^2$, then since $-P = \sigma_U(c^2/\mathbf{R})$, density diminishes as $1/\mathbf{R}$ as deduced above;

$$\rho_U = 3(\sigma_U)/\mathbf{R} \tag{34}$$

The above relationships can now be applied to explain the expansion profile of the universe in terms of the critical condition that causes masses to be gravitationally detached. Thus, for a Hubble sphere having an effective radius R and density ρ_U, the gravitational force will be exceeded by the exzansion acceleration balanced against the expansion acceleration when:

$$\frac{\mathbf{M}_U\mathbf{M}_b G}{\mathbf{r}^2} < \frac{\mathbf{M}_b c^2}{\mathbf{R}} \tag{35}$$

Where $\mathbf{M}_u$ is the operative mass of the universe, and $\mathbf{M}_b$ is the mass of a body at distance '$\mathbf{r}$' from the Hubble center. Substituting for $\rho_U(4\pi\mathbf{r}^3/3)$ for $\mathbf{M}_U$ and $c^2/4\pi\mathbf{R}\sigma_U$, (35) becomes:

Figure 4

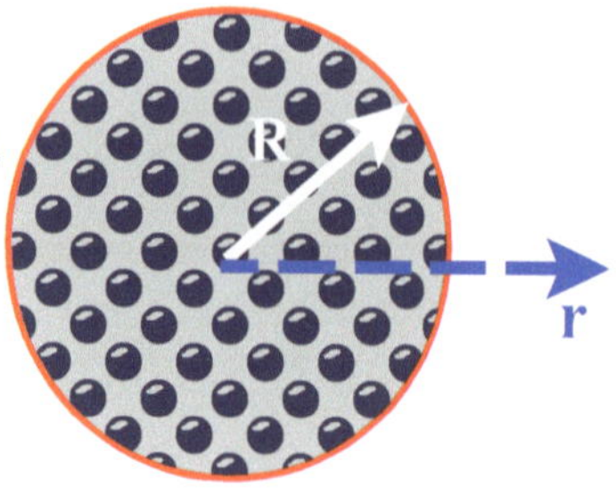

$$\frac{4\pi\rho_U r^3 M_b(c^2)}{3(4\pi)(r^2)R\sigma_U} < \frac{M_b(c^2)}{R} \qquad (36)$$

And since $\rho_U = 3\sigma_U/R$, then:

$$R < r \qquad (37)$$

At distance 'r' > R, spatial recession exceeds Hubble dilation (dR/dt). Viewed from the Hubble center and biased by the presumption that all receding nebula are comoving with recessional space, the nebula appear to get a boost. Photons reaching the earth (emitted in the direction of the earth from nebula beyond the Hubble sphere at the time they were emitted), are interpreted as evidence of an onset in the rate of cosmological expansion.[48] Of significance, the equations [e.g, (16)] derived therefrom, are consistent with the notion of past eternal expansion. The inverse dependence of G and ρ_U upon R cancel -- consequently the doctrine of acquired inertia predicts an expansion profile that comports with Λ.

Dynamic space, however, does not require additional postulation to explain gravity. Static curvature does not exist because space is not static. Ironically, the cosmolical constant Λ does not oppose gravity, it is the root cause of gravity. Without expansion, there is no gravity, and without gravity there is no gravitational collapse. Two profound mysteries are at once resolved:

1) The ultimate fate of the universe: In the end, there is no end, at least by means of gravitational collapse.

2) The fine tuned balance between expansion and gravity: A perception resulting from the erroneous belief that gravity and expansion are independent phenomena.

Einstein's $\Lambda = (3H^2)$, unknowingly unlocked the root cause of gravity. G gets demoted from a fundamental constant of the universe to a pseudo force created acceleration that diminishes with age. Spatial acceleration ($3H^2R$) is subliminally encoded within G per (10).

Both Newton and Einstein opined that gravity must be the result of an ongoing action.[49] The nature of this subtlety as a pseudo force, (footnote 4) was first pondered by Richard Feynman.[50]

---

[48] This has prompted an extensive search for a latent form of dark energy tailored by hypothesis to have the behavioral complexity that fits the rational.

[49] *"Gravity must be caused by an agent acting constantly, according to certain laws, but whether this agent be material or immaterial I have left to the consideration of my readers"* (Isaac Newton).

[50] *"One very important feature of pseudo forces is that they are always proportional to the masses. The same is true of gravity. The possibility exists therefore that gravity itself is a pseudo force. Is it not possible that perhaps gravitation is due simply to the fact we do not have the right coordinate system."* [Feynman, Lectures On Physics, Vol 1 at 12-11].

As further developed herein *a la* (21) and the discussion pertaining thereto, the gravitational frame that alluded discovery is the isotropically expanding rest frame of the universe. More particularly, each and every point in free space expands equally in all directions. 2nd law symmetry symbolically:

$$F_{ma} = \mathbf{V}_U(\ddot{\mathbf{M}}) < ----- > \mathbf{M}(\ddot{\mathbf{V}}_U) = F_G \tag{38}$$

In words, (38) symbolizes the equivalence between 2nd law forces. Force is created irrespective of whether space or mass accelerates. In both cases, the functional contribution of the Hubble universe is defined by the infinite inertial plane σ_U. For a spherical mass $\mathbf{M}$ undergoing unidirectional acceleration, force is distributed along the line of action opposite to the direction of acceleration taking into account the lineal mass In **Figure 5A** the infinite plane [orange], depicted flat and orthogonal to the direction of acceleration, exerts a nonuniform reactionary counter field. **Figure 5B** illustrates the radially convergent reactionary force for an isotropically divergent spatial expansion field. The imaginary inertial plane must at all points be orthogonal to the primary line of action to issue an oppositely directed counter reaction, requiring σ_U operative exist as a spherical shell (mathematically an infinite number of segments from an infinite number of infinite planes (blue) sewn together to effect a shell that issues counter forces (red). The counter pressure $\mathbf{g}\sigma_U$ will balance the pressure created by the expansion acceleration ($\mathbf{c^2/r}$) acting upon the contribution of the all mass colinear with the line of action. For analytical purposes then, the criteria for balanced pressure is satisfied if σ_U is placed in contact with the surface of $\mathbf{M}$ as shown in **Figure 5C**.

That spatial divergence within the volume surrounded by the surface can be expressed as the integral of spatial flux normal to the surface, allows placement of the **kgm/meter²** inertial shell in contact with the surface of the volume undergoing spatial expansion. Again earth as example:

$$\frac{F}{A} = \frac{c^2}{R} \times \frac{M_E}{4\pi(r)^2} = \frac{9 \times 10^{16}}{1.08 \times 10^{26}} \times \frac{5..98 \times 10^{24}}{12.56 \times (6.37 \times 10^6)^2} \approx \frac{9.8\ ntn}{m^2} \tag{39}$$

A surrounding contiguous shell having a density of one **kg** per **m²**, will impart a reactionary force :

$$\frac{Force}{kg} = \frac{F}{A} \times \frac{m^2}{kg} = \frac{9.8\ ntn}{m^2} \times \frac{m^2}{kg} = \frac{9.8\ ntn}{kgm} \tag{40}$$

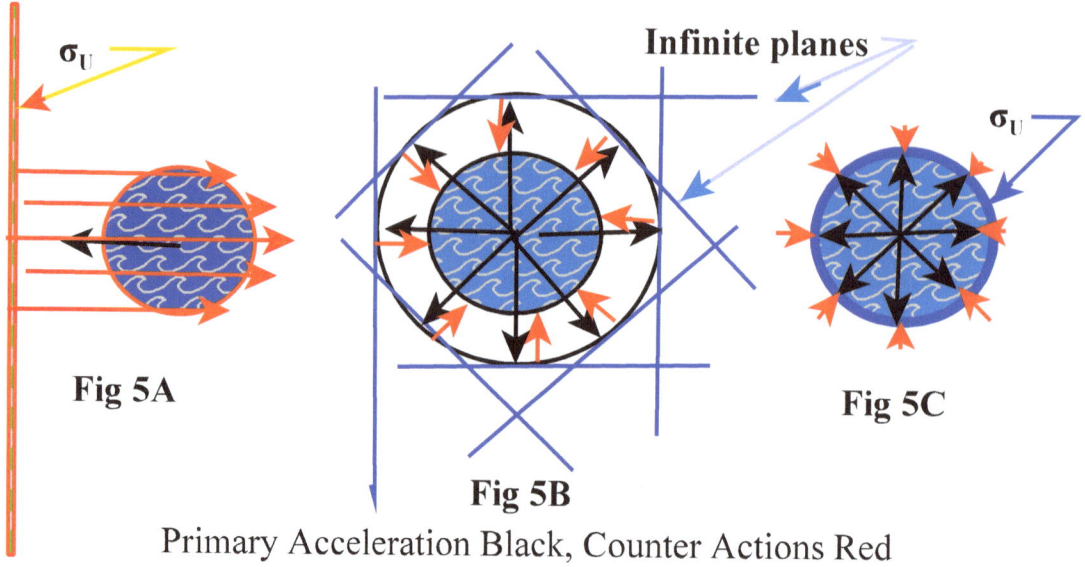

σ_U   Infinite planes   σ_U

Fig 5A

Fig 5C

Fig 5B

Primary Acceleration Black, Counter Actions Red

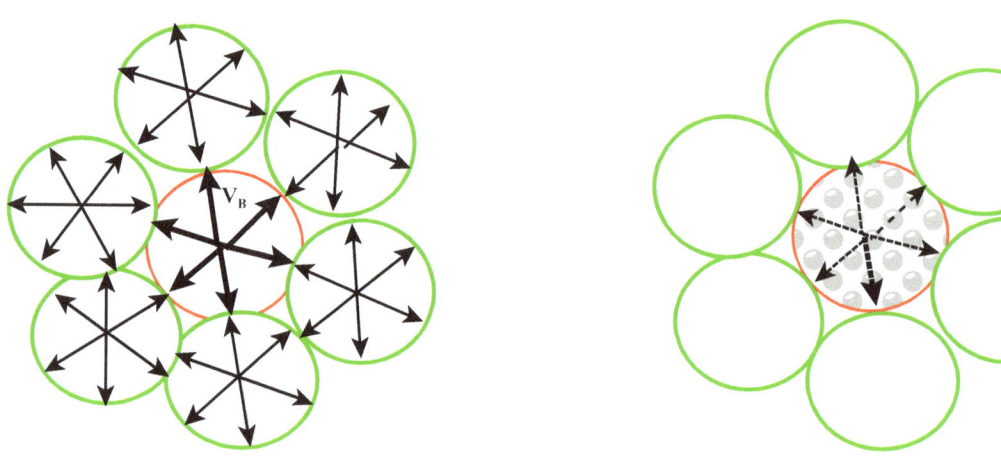

Figure 6A   **Figure 6B**

Shown in Figure 6A, an expanding empty volume V_B enclosed by a surface **S** (red) of area A_B is surrounded by six identical empty volumes (green). All are expanding at a rate $a_n = c^2/R$ as indicated by the solid black arrows. In Figure 6B, mass has been is sprinkled throughout the central volume V_B to create a uniform density ρ_B. The total mass enclosed by **S** is therefore:

$$M_B = V_B(\rho_B). \qquad (41)$$

The outward pressure exerted by the **S** surface due to the enclosed mass M_B is consequently reduced by the factor:

$$V_B(\rho_B)(a_n)/A_B \qquad (42)$$

Figure 6B, the reduced pressure (dotted black arrows) resulting from the addition of mass M_B to V_B is offset by the influx of momentum from expansion of the six surrounding volumes. Figure 6C.

This causes the surrounding green volumes to fill the pressure deficit (Figure 6C) and the next level of expanding volumes (blue) fill the void left by displaced green volumes, and so forth, the influx momentum (Pressure) crossing the **S** surface being the product of the inertial operative σ_U multiplied by the gravitational convergence 'g' created by spatial expansion acting upon $\mathbf{M_B}$ as per equation (4). The influx from the universe divided by the area of S is the gravity field of $\mathbf{M_B}$.

$$g = [\sigma_B/(\sigma_U)](a_n) \tag{43}$$

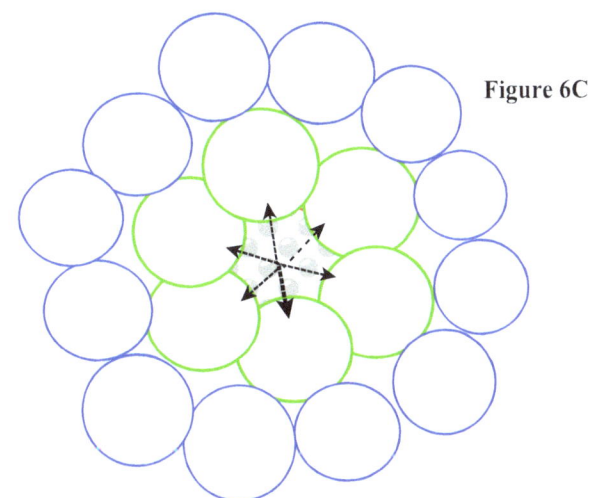

Figure 6C

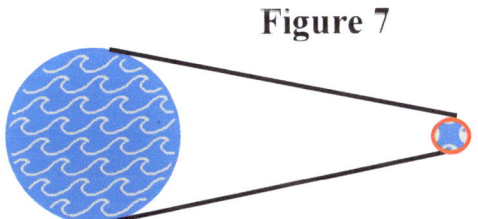

Figure 7

Figure 7 depicts the earth shrunk to the size of a black hole. (about one centimeter in radius). Would the intensity of the **g** field at the surface of an imaginary shell having radius equal to the earth be unchanged? In general, density has no effect upon the 'g' force -- force is independent of the volume occupied by the mass[51] -- equations (39) and (40) depend only upon mass and the area over which the field is distributed. Ergo, gravitational phenomena is not dependent upon expansion of space within the volume enclosed by the surface over which the intensity of the field is predicated Masses simply create local inertial reactions that manifest as surface pressure. To the degree accelerated mass creates collinear displacement,

---

[51]When the earth or any other body is compressed to the size of a black hole, there is a contribution from the action of gravity acting upon gravity as discussed in connection with equation 37). For a BH created from ordinary bare mass using the conventional formula, the additional gravity field of gravity acting upon gravity will increase the intensity by approximately 25% depending upon how the black hole is modeled.

isotropic spatial expansion creates gravitational convergence flux. The absence of a freely expanding spatial volume (in form as condensed matter) is to neighboring places a Pascal's law demand upon the universe, as a whole, to equalize pressure -- '**g**' fields, although proportional to mass, are in reality, convergent momentum flows -- the continuous response provoked by the action of the expansion field in concert with the infinite inertial plane σ_U.[52] This is as it must be to satisfy momentum conservation in an incompletely homogenized (lumpy universe). The decisive referent is obtained by equating Newtonian gravitational acceleration [MG/r^2] with expansion acceleration c^2/R at the limit of the Hubble scale, whence $R = r$, and $M = M_U$, then:

$$M_U G = Rc^2 \tag{44}$$

That (44) reveals the physiology underlying the mystery ratio (19), follows from the alternative but fully equivalent expression per (10):

$$G = Rc^2/M_U \tag{45}$$

That (10) and (45) are identical, it is only necessary to substitute $4\pi R^2 \sigma_U$ for M_U from (32).

---

[52]"*Gravity must be caused by an agent acting constantly according to certain laws, but whether this agent be material or immaterial, I have left to the consideration of my readers.*" (Written by Isaac Newton in his 4th letter to Richard Bentley circa 1692.

A compendium of Short Answers
To Interesting Question

A brief history of Lambda

The universe is a work in process, requiring flexibility between inertia, space, time, gravity and expansion to account for the evolving state of it's defining parameters. To assign fixed values to **G** and **M** motivated by doubtful theories, leads to compounded hypothesis inflexible formalisms.

The formalization of gravity as mass induced curvature of static space was later discredited by the discovery of expansion. It is somewhat surprising that Einstein did not then adapt the cosmological constant to explain gravity in terms of dynamic space (Having already copped the ΛR correctly as (c^2/R), the prediction of **G** follows straightaway). But the perception of gravity as a reactionary force would await the musings of Richard Feynman some 40 years later. In this lost opportunity, it is worth noting that the entire physics community failed to make the connection. It was not until the 1998 supernova studies, that cosmologists began to rethink the significance of Λ.

The *"Lambda-Cold-Dark-Matter"* model of the universe is now *"Standard Theory."* And while the recognition of Λ as exponentially expanding space is fitting and proper, the (ΛCDM) model misstates its origin. The cosmological constant is not the result of hypothetical *"dark energy,"* rather the positive mass energy of the universe depends upon expansion.

Of the many drawn to the mystery of gravity, it was Richard Feynman that first opined that "gravity" could be a pseudo force. Today, the (ΛCDM) model has not been applied to derive **G**, nor do the guardians of standard recognize gravity as a "pseudo force." But it is not promulgators of standard theory that determine the operation of the universe. It is the underlying equations upon which the laws of physics are founded. In the last analysis, gravity and expansion are interdependent and it is the latter that powers the universe.

The search for gravitational mechanism proceeds with the hunt for dark energy. In the meantime, space continues to expand, inertia continues to augment, and gravity continues diminish, all of which shall remain a mystery to those who, in the past, have always been wrong, but never in doubt.

Rest Mass is Negative Energy?

It's straightforward to see that for a body of mass **M** moving at velocity **v**, momentum is **mv** and it's easy to understand that a body moving at velocity **v** has kinetic energy **(Mv²)/2**. But why should the mass of a body "at rest" be expressible in terms of motion? What is it that is moving. One might guess it is the electrons in matter, but that would be incomplete and incorrect. Einstein wrote several papers attempting to find a general proof of **E =Mc²** that did not involve light and/or radiation, but was unsuccessful.[53]

Here is an answer conjured by the Author some years past. It follows from a now famous assertion by Richard Feynman: *"It costs nothing to create a mass at the center of the universe"* Since every point is the center of its own Hubble sphere, the statement applies universally. It has become a fundament of the *"Zero Energy Theory of the Cosmos."* The question posed is that of how much energy is required to transport a mass **M** from the edge of a Hubble sphere to its center while moving against the recessional flow of space?

For an exponentially accelerating universe, expansion rate is **(c²)/R**. The force **F** is:

$$\mathbf{F} = \mathbf{Ma} \ = \ \mathbf{M[(c^2)/R]}$$

and the energy is **E** is

$$\mathbf{E} = \int \mathbf{F} \cdot \mathbf{ds} = \int_R^0 \mathbf{M}\left[\frac{\mathbf{c}^2}{\mathbf{R}}\right]\mathbf{dr} = \frac{\mathbf{Mc}^2}{\mathbf{R}}\int_R^0 \mathbf{dr} = -\mathbf{Mc}^2$$

Total energy to move a body of mass **M** from the Hubble edge to its center is (**-Mc²**). This corresponds to the negative energy in the '**g**' field of a mass **M**. Net cosmic energy is zero per Feynman.

53